Novel Algorithms for Fast Statistical Analysis of Scaled Circuits

Lecture Notes in Electrical Engineering
Volume 46

For other titles published in this series, go to
www.springer.com/series/7818

Amith Singhee • Rob A. Rutenbar

Novel Algorithms for Fast Statistical Analysis of Scaled Circuits

Dr. Amith Singhee
IBM Corporation
T. J. Watson Research Center
1101 Kitchawan Road
Route 134
PO Box 218
Yorktown Heights, NY 10598
USA
asinghee@us.ibm.com

Rob A. Rutenbar
Carnegie Mellon University
Dept. Electrical & Computer Engineering
5000 Forbes Ave.
Pittsburg, PA 15213-3890
USA
rutenbar@ece.cmu.edu

ISSN 1876-1100 Lecture Notes in Electrical Engineering
ISBN 978-90-481-3099-3   e-ISBN 978-90-481-3100-6
DOI 10.1007/978-90-481-3100-6
Springer Dordrecht Heidelberg London New York

Library of Congress Control Number: 2009931791

© Springer Science + Business Media B.V. 2009
No part of this work may be reproduced, stored in a retrieval system, or transmitted in any form or by any means, electronic, mechanical, photocopying, microfilming, recording or otherwise, without written permission from the Publisher, with the exception of any material supplied specifically for the purpose of being entered and executed on a computer system, for exclusive use by the purchaser of the work.

Printed on acid-free paper

Springer is part of Springer Science+Business Media (www.springer.com)

To my parents
– Amith

Introduction

I.1 Background and Motivation

Very Large Scale Integration (VLSI) technology is moving deep into the nanometer regime, with transistor feature sizes of 45 nm already in widespread production. Computer-aided design (CAD) tools have traditionally kept up with the difficult requirements for handling complex physical effects and multi-million-transistor designs, under the assumption of fixed or deterministic circuit parameters. However, at such small feature sizes, even small variations due to inaccuracies in the manufacturing process can cause large relative variations in the behavior of the circuit. Such variations may be classified into two broad categories, based on the source of variation: (1) *systematic* variation, and (2) *random* variation. Systematic variation constitutes the deterministic part of these variations; e.g., proximity-based lithography effects, nonlinear etching effects, etc. [GH04]. These are typically pattern dependent and can potentially be completely explained by using more accurate models of the process. Random variations constitute the unexplained part of the manufacturing variations, and show stochastic behavior; e.g., gate oxide thickness (t_{ox}) variations, poly-Si random crystal orientation (RCO) and random dopant fluctuation (RDF) [HIE03]. These random variations cannot simply be accounted for by more accurate models of the physics of the process because of their inherent random nature (until we understand and model the physics well enough to accurately predict the behavior of each ion implanted into the wafer).

As a result, integrated circuit (IC) designers and manufacturers are facing difficult challenges in producing reliable high-performance circuits. Apart from the sheer size and complexity of the design problems, a relatively new and particularly difficult problem is that of these *para-*

metric variations (threshold voltage (V_t), gate oxide thickness, etc.) in circuits, due to nonsystematic variations in the manufacturing process. For older technologies, designers could afford to either ignore the problem, or simplify it and do a worst-case corner based conservative design. At worst, they might have to do a re-spin to bring up the circuit yield. With large variations, this strategy is no longer efficient since the number of re-spins required for convergence can be prohibitively large. *Per-transistor* effects like RDF and line edge roughness (LER) [HIE03] are becoming dominant as the transistor size is shrinking. As a result, the relevant statistical process parameters are no longer a few inter-wafer or even inter-die parameters, but a huge number of inter-device (intra-die) parameters. Hence, the dimensionality with which we must contend is also very large, easily 100s for custom circuits and millions for chip-level designs. Furthermore, all of these inter-die and intra-die parameters can have complex correlation amongst each other. Doing a simplistic conservative design will, in the best case, be extremely expensive, and in the worst case, impossible. These variations must be modeled accurately and their impact on the circuit must be predicted reliably in most, if not all, stages of the design cycle. These problems and needs have been widely acknowledged even amongst the non-research community, as evidenced by this extensive article [Ren03].

Many of the electronic design automation (EDA) tools for modeling and simulating circuit behavior are unable to accurately model and predict the large impact of process-induced variations on circuit behavior. Most attempts at addressing this issue are either too simplistic, fraught with no-longer-realistic assumptions (like linear [CYMSC85] or quadratic behavior [YKHT87][LLPS05], or small variations), or focus on just one specific problem (e.g., Statistical Static Timing Analysis or SSTA [CS05][VRK+04a]). This philosophy of doing "as little as needed", which used to work for old technology nodes, will start to fail for tomorrow's scaled circuits. There is a dire need for tools that efficiently model and predict circuit behavior in the presence of large process variations, to enable reliable and efficient design exploration. In the cases where there are robust tools available (e.g., Monte Carlo simulation [Gla04]), they have not kept up with the speed and accuracy requirements of today's, and tomorrow's, IC variation related problems.

In this thesis we propose a set of novel algorithms that discard simplifications and assumptions as much as possible and yet achieve the necessary accuracy at very reasonable computational costs. We recognize that these variations follow complex statistics and use *statistical* approaches based on accurate statistical models. Apart from being flexible and scalable enough to work for the expected large variations in future VLSI technologies, these techniques also have the virtue of being independent of the problem domain: they can be applied to any engineering or scientific problem of a similar nature. In the next section we briefly review the specific problems targeted in this thesis and the solutions proposed.

I.2 Major Contributions

In this thesis, we have taken a wide-angle view of the issues mentioned in the previous section, addressing a variety of problems that are related, yet complementary. Three such problems have been identified, given their high relevance in the nanometer regime; these are as follows.

I.2.0.1 SiLVR: Nonlinear Response Surface Modeling and Dimensionality Reduction

In certain situations, SPICE-level circuit simulation may not be desired or required, for example while computing approximate yield estimates inside a circuit optimization loop [YKHT87][LGXP04]: circuit simulation is too slow in this case and we might be willing to sacrifice some accuracy to gain speed. In such cases, a common approach is to build a model of the relationship between the statistical circuit parameters and the circuit performances. This model is, by requirement, much faster to evaluate than running a SPICE-level simulation. The common term employed for such models is *response surface models* (RSMs). In certain other cases, we may be interested in building an RSM to extract specific information regarding the circuit behavior, for example, sensitivities of the circuit performance to the different circuit parameters. Typical RSM methods have often made simplifying assumptions regarding the characteristics of the relationship being modeled (e.g., linear behavior [CYMSC85]), and have been sufficiently accurate in the past. However, in scaled technologies, the large extent and number of variations make these assumptions invalid.

In this thesis, we propose a new RSM method called *SiLVR* that discards many of these assumptions and is able to handle the problems posed by highly scaled circuits. SiLVR employs the basic philosophy of *latent variable regression*, that has been widely used for building linear models in chemometrics [BVM96], but extends it to flexible nonlinear models. This model construction philosophy is also known as *projection pursuit*, primarily in the statistics community [Hub85]. We show how SiLVR can be used not only for performance modeling, but also for extracting sensitivities in a nonlinear sense and for output-driven dimensionality reduction from 10–100 dimensions to 1–2. The ability to extract insight regarding the circuit behavior in terms of numerical quantities, even in the presence of strong nonlinearity and large dimensionality, is the real strength of SiLVR. We test SiLVR on different analog and digital circuits and show how it is much more flexible than state-of-the-art quadratic models, and succeeds even in cases where the latter completely breaks down. These initial results have been published in [SR07a].

I.2.0.2 Fast Monte Carlo Simulation Using Quasi-Monte Carlo

Monte Carlo simulation has been widely used for simulating the statistical behavior of circuit performances and verifying circuit yield and failure probability [HLT83], in particular for custom-designed circuits like analog circuits and memory cells. In the nanometer regime, it will remain a vital tool in the hands of designers for accurately predicting the statistics of manufactured ICs: it is extremely flexible, robust and scalable to a large number of statistical parameters, and it allows arbitrary accuracy, of course at the cost of simulation time. In spite of the technique having found widespread use in the design community, it has not received the amount of research effort from the EDA community that it deserves. Recent developments in number theory and algebraic geometry [Nie88][Nie98] have brought forth new techniques in the form of *quasi-Monte Carlo*, which have found wide application in computational finance [Gla04][ABG98][NT96a]. In this thesis, we show how we can significantly speed up Monte Carlo simulation-based statistical analysis of circuits using quasi-Monte Carlo. We see speedups of $2\times$ to $50\times$ over standard Monte Carlo simulation across a variety of transistor-level circuits. We also see that quasi-Monte Carlo scales better in terms of accuracy: the speedups are bigger for higher accuracy requirements. These initial results were published in [SR07b].

I.2.0.3 Statistical Blockade: Estimating Rare Event Statistics, with Application to High Replication Circuits

Certain small circuits have millions of identical instances on the same chip, for example, the SRAM (Static Random Access Memory) cell. We term this class of circuits as *high-replication circuits*. For these circuits, typical acceptable failure probabilities are extremely small: orders of magnitude less than even 1 part-per-million. Here we are restricting ourselves to failures due to parametric manufacturing variations. Estimating the statistics of failures for such a design can be prohibitively slow, since only one out of a million Monte Carlo points might fail: we might need to run millions to billions of simulations to be able to estimate the statistics of these very rare failure events. Memory designers have often avoided this problem by using analytical models, where available, or by making

Introduction xi

"educated guesses" for the yield, using large safety margins, worst-case corner analysis, or small Monte Carlo runs. Inaccurate estimation of the circuit yield can result in significant numbers of re-spins if the margins are not sufficient, or unnecessary and expensive (in terms of power or chip area) over-design if the margins are too conservative. In this thesis, we propose a new framework that allows fast sampling of these rare failure events and generates analytical probability distribution models for the statistics of these rare events. This framework is termed *statistical blockade*, inspired by its mechanics. Statistical blockade brings down the number of required Monte Carlo simulations from millions to very manageable thousands. It combines concepts from machine learning [HTF01] and extreme value theory [EKM97] to provide a novel and useful solution for this under-addressed, but important problem. These initial results have been published in [SR07c][WSRC07][SWCR08].

I.3 Preliminaries

A few conventions that will be followed throughout the thesis are worth mentioning at this stage. Each statistical parameter will be modeled as having a probability distribution that has been extracted and is ready for use by the algorithms proposed in this thesis. The parameters considered are SPICE model parameters, including threshold voltage (V_t) variation, gate oxide thickness (t_{ox}) variation, resistor value variation, capacitor value variation, etc. It will be assumed for experimental setup, that the statistics of any variation at a more physical level, e.g., random dopant fluctuation, can be modeled by these probability distributions of the SPICE-level device parameters.

Some other conventions that will be followed are as follows.

- All vector-valued variables will be denoted by bold small letters, for example $\mathbf{x} = \{x_1, \ldots, x_s\}$ is a vector in s-dimensional space with s coordinate values, also called an s-vector. Rare deviations from this rule will be specifically noted. Scalar-valued variables will be denoted with regular (not bold) letters, and matrices with bold capital letters; for example, $\mathbf{X}$ is a matrix, where the i-th row of the matrix is a vector $\mathbf{x}_i$. All vectors will be assumed to be column vectors, unless transposed. $\mathbf{I}_s$ will be the $s \times s$ identity matrix.

- We will use s to denote the dimensionality of the statistical parameter space that any proposed algorithm will work in.

- Following standard notation, $\mathbb{R}$ denotes the set of all real numbers, $\mathbb{Z}$ denotes the set of all integers, $\mathbb{Z}_+$ denotes the set of all nonnegative integers, and $\mathbb{R}^s$ is the s-dimensional space of all real-valued s-vectors.

I.4 Organization

The ideas proposed in this thesis are born out of a large body of knowledge from several different fields. Hence, there is no practical limit to the amount of background material that could be considered relevant. It is out of the practical scope of any single volume to cover all such "relevant" material in detail. However, to make these ideas accessible to the general reader, a reasonably comprehensive discussion of the background is needed. In its attempt to achieve a balance between conciseness and completeness, this thesis reviews relevant background material that is required for a clear understanding of the proposed ideas, and avoids lengthy expositions of background material on related or competing ideas. The latter can easily be found in referenced literature in, or related to, electronic design automation, and is not immediately required for a clear understanding of the proposed ideas. In certain cases, small diversions are made to review interesting concepts from some field outside of electrical and computer engineering, to enable a more expansive understanding of the underlying concepts. An example is the brief review of Asian option pricing in Sect. 2.2.1.1.

This thesis is organized into three nearly independent chapters, each presenting one of the three contributions of this work. Chapter 1 introduces SiLVR, the proposed nonlinear RSM method. For this purpose, it first reviews typical RSM techniques and relevant background relating to latent variable regression, projection pursuit, and the specific techniques employed by SiLVR. The chapter ends with a section comparing the modeling results of SiLVR against simulation and an optimal quadratic RSM (PROBE from [LLPS05]). Chapter 2 provides the necessary application and theoretical background for Monte Carlo simulation and the proposed quasi-Monte Carlo (QMC) simulation technique. It then details the proposed QMC flow and present experimental results validating its gains over standard Monte Carlo. Chapter 3 introduces the problem of yield estimation for high-replication circuits and reviews relevant background from machine learning and extreme value theory. It then explains the proposed statistical blockade flow in detail and present validation using different relevant circuit examples. Chapter 4 provides concluding remarks. Suggestions for future research directions are provided at the end of each of Chaps. 1, 2 and 3.

Contents

1. SiLVR: Projection Pursuit for Response Surface Modeling   1
   - 1.1 Motivation   1
   - 1.2 Prevailing Response Surface Models   4
     - 1.2.1 Linear Model   4
     - 1.2.2 Quadratic Model   5
     - 1.2.3 PROjection Based Extraction (PROBE): A Reduced-Rank Quadratic Model   6
   - 1.3 Latent Variables and Ridge Functions   8
     - 1.3.1 Latent Variable Regression   8
     - 1.3.2 Ridge Functions and Projection Pursuit Regression   10
   - 1.4 Approximation Using Ridge Functions: Density and Degree of Approximation   13
     - 1.4.1 Density: What Can Ridge Functions Approximate?   14
     - 1.4.2 Degree of Approximation: How Good Are Ridge Functions?   16
   - 1.5 Projection Pursuit Regression   18
     - 1.5.1 Smoothing and the Bias–Variance Tradeoff   19
     - 1.5.2 Convergence of Projection Pursuit Regression   21
   - 1.6 SiLVR   27
     - 1.6.1 The Model   27
     - 1.6.2 On the Convergence of SiLVR   31
     - 1.6.3 Interpreting the SiLVR Model   33
     - 1.6.4 Training SiLVR   36

	1.7	Experimental Results	44
		1.7.1 Master–Slave Flip-Flop with Scan Chain	45
		1.7.2 Two-Stage RC-Compensated Opamp	47
		1.7.3 Sub-1 V CMOS Bandgap Voltage Reference	52
	1.8	Future Work	55
2.	Quasi-Monte Carlo for Fast Statistical Simulation of Circuits		59
	2.1	Motivation	59
	2.2	Standard Monte Carlo	61
		2.2.1 The Problem: Bridging Computational Finance and Circuit Design	61
		2.2.2 Monte Carlo for Numerical Integration: Some Convergence Results	64
		2.2.3 Discrepancy: Uniformity and Integration Error	67
	2.3	Low-Discrepancy Sequences	72
		2.3.1 (t, m, s)-Nets and (t, s)-Sequences in Base b	72
		2.3.2 Constructing Low-Discrepancy Sequences: The Digital Method	76
		2.3.3 The Sobol' Sequence	82
		2.3.4 Latin Hypercube Sampling	88
	2.4	Quasi-Monte Carlo in High Dimensions	92
		2.4.1 Effective Dimension of the Integrand	94
		2.4.2 Why Is Quasi-Monte Carlo (Sobol' Points) Better Than Latin Hypercube Sampling?	98
	2.5	Quasi-Monte Carlo for Circuits	101
		2.5.1 The Proposed Flow	101
		2.5.2 Estimating Integration Error	103
		2.5.3 Scrambled Digital (t, m, s)-Nets and (t, s)-Sequences	106
	2.6	Experimental Results	108
		2.6.1 Comparing LHS and QMC (Sobol' Points)	109
		2.6.2 Experiments on Circuit Benchmarks	113
	2.7	Future Work	121
3.	Statistical Blockade: Estimating Rare Event Statistics		123
	3.1	Motivation	123
	3.2	Modeling Rare Event Statistics	126

	3.2.1	The Problem	126
	3.2.2	Extreme Value Theory: Tail Distributions	128
	3.2.3	Tail Regularity Conditions Required for $F \in \mathrm{MDA}(H_\xi)$	131
	3.2.4	Estimating the Tail: Fitting the GPD to Data	133
3.3	Statistical Blockade		137
	3.3.1	Classification	137
	3.3.2	Support Vector Classifier	138
	3.3.3	The Statistical Blockade Algorithm	142
	3.3.4	Experimental Results	145
3.4	Making Statistical Blockade Practical		155
	3.4.1	Conditionals and Disjoint Tail Regions	155
	3.4.2	Extremely Rare Events and Statistics	159
	3.4.3	A Recursive Formulation of Statistical Blockade	163
	3.4.4	Experimental Results	166
3.5	Future Work		169

4. Concluding Observations — 171

Appendices — 175

Appendix A Derivations of Variance Values for Test Functions in Sect. 2.6.1 — 175

    A.1 Variance of f_c — 175

    A.2 One Dimensional Variance of f_s — 178

References — 181

Index — 193

Chapter 1
SiLVR: Projection Pursuit for Response Surface Modeling

1.1 Motivation

In many situations it is desirable to have available an inexpensive model for predicting circuit performance, given the values of various statistical parameters in the circuit (e.g., V_t for the different devices in the circuit). Examples of such situations are 1) in a circuit optimization loop where quick estimates of yield might be necessary to drive the solution towards a high-yield design in reasonable run time, and 2) during manual design, a simple analytical model can provide insight into circuit operation using metrics such as sensitivities or using quick visualization, thus helping the designer to understand and tune the circuit. [DFK93] provides a good overview of general statistical design approaches. Even though the paper is not very recent, much of the literature on statistical design (yield optimization) over the last couple of decades proposes techniques that fall under the general types discussed therein. Such performance models in the statistical parameter space are commonly referred to as *response surface models*: we abbreviate this as RSM in this thesis. Initial approaches employed linear regression to model circuit performance metrics, as in [CYMSC85]. Soon, the linear models were found to be inadequate for modeling nonlinear behavior and quadratic models were proposed in [YKHT87][FD93] to reduce the modeling error.

These low-order models worked sufficiently well for the technologies of yesteryears, but face fundamental difficulties going forward. Any solution now must address three large challenges:

- **Dimensionality**: The number of sources of variations in the circuit can be large. Even for a simple flip-flop, there can be over 50 sources, e.g., random dopant fluctuation (RDF), line edge roughness

(LER), random poly crystal orientation (RCO) [HIE03], and gate oxide thickness variation. Although many such sources can be absorbed into a few device-level parameters, for larger analog cells the dimensionality can still easily be in the hundreds. The number of variables, s, in a model determines the number of unknown model parameters that need to be estimated during model fitting. The number of SPICE-simulated points needed then is at least the number of unknown parameters.

- **Large variations**: The relative effect of every variation source is becoming very large. Just considering RDF, predictions indicate that the standard deviation of V_t can be 10% of the nominal V_t at the 70 nm node [EIH02], growing to 21% for a 25 nm device [FTIW99], with 0.3 V V_t. If other variations (LER, RCO) are considered, the deviation is even higher [HIE03].

- **Nonlinearity**: Not all performance/variable relationships are simple. A good example is the relationship between device V_t in a flip-flop, and the flip-flop delay. Such nonlinearity is even more pronounced in the case of analog circuits.

Linear models are able to handle the dimensionality well, since the number of unknown model parameters is a slowly increasing $s+1$, where s is the number of input variables. These models, however, fail to capture nonlinear behaviors, for which higher-order models are needed. Higher-order models can, however, have a very large number of parameters: a polynomial of degree d in s dimensions has $^{s+d}C_d$ terms. Hence, even a quadratic model in 100 dimensions can have 5,151 parameters, requiring 5,151 initial SPICE simulations to generate training points for the model! Recent attempts at reducing the number of unknowns in the quadratic model have resulted in very efficient techniques, namely PROBE [LLPS05] and kernel reduced rank regression (RRR) [FL06]. Both these methods essentially reduce the rank of the quadratic model, the former doing it in a more natural near-optimal manner. We will look at PROBE in more detail in Sect. 1.2.3. However, these methods still suffer from the severe restriction of quadratic (which includes linear) behavior. In the presence of large variations, the nonlinearity in the circuit behavior is significant enough to make these models unusable, as we shall see in later sections.

In this chapter, we review the *latent variable regression* (LVR) [BVM96] and *projection pursuit regression* (PPR) [Hub85] strategies and show why they can be attractive in these scenarios. Roughly speaking, these techniques iteratively extract the next statistically most important variable (latent variable or LV), and minimize the error in fitting the remainder of the unexplained performance variation. Hence, they directly reduce the problem dimensionality. Further, these techniques can be accompanied with flexible, but compact, functional forms for the model, thus reducing a priori assumptions about the magnitude of variations and the behavior modeled. Using these ideas this chapter will develop an RSM strategy for silicon design problems – SiLVR – and show its superior performance in comparison to PROBE, in the context of the three challenges mentioned above. We will also see how the "designer's insight" can be obtained naturally from the structure of the SiLVR model, in the form of some quantitative measures and insightful visualization. Such insights into the circuit behavior can help the designer to better understand the behavior of the circuit during manual design, and guide the optimizer better during automatic sizing. SiLVR was first introduced by us in [SR07a].

LVR methods have a long and interesting track record and encompass a variety of different techniques that follow the same philosophy to meet slightly different objectives; for example, partial least squares (PLS) [WRWI84], canonical correlation regression (CCR) [BVM96], and reduced rank regression [RV98]. A good survey and comparison is provided in [BVM96]. These LVR techniques have found wide application, much of it outside the realm of silicon application, in areas ranging from chemometrics [WSE01] to statistics [DT82], to bioinformatics [BS06]. Many LVR methods still assume a linear relationship, or use a low-order nonlinear kernel to explain the assumed nonlinear relationships. Thus, our own interest is in LVR methods that support a more flexible nonlinear framework. Here, Baffi et al. [BMM99] and Malthouse et al. [MTM97] are noteworthy. In addition to the single variable iterative extraction philosophy, these show how to use a neural network [Rip96] to capture significant nonlinear behaviors. However, [BMM99] suffers from unreliable and slow convergence during training and [MTM97] uses an excessively complex model that can be prone to computational burdens and overfitting. SiLVR, although similar in flavor to these methods, uses a more compact model in a new unified training framework to remove these issues.

Although SiLVR derives its name from LVR, its philosophy finds a closer fit with projection pursuit [FS81][Hub85]. Both LVR and PP are very similar in the way they operate, but their theory and applications seem to have developed more or less independently: LVR in the world of chemometrics (PLS) and statistics (CCR, RRR), while PP in the world of statistics, approximation theory and machine learning. Theoretical foundations for PP appear to be better developed, more so for nonlinear regression and the particular case of the SiLVR model (PP using sigmoidal functions). We will review relevant results from these as we move toward developing the SiLVR model architecture.

In the rest of this chapter, we briefly review linear and quadratic models, including PROBE, a low-rank quadratic model, after which we review the LVR and PP techniques, along with relevant theoretical results from approximation theory. Finally, we develop the SiLVR model, covering relevant details regarding model training, and show experimental results.

1.2 Prevailing Response Surface Models

Before we review linear and quadratic models, let us first concretely define the RSM problem. Let $\mathcal{X} = \mathbb{R}^s$ be the statistical parameter space and $\mathcal{Y} = \mathbb{R}^{s_y}$ be the circuit performance-metric or output space: s_y is the number of outputs. For a given $\mathbf{x} \in \mathcal{X}$, $\mathbf{y} = \mathbf{f}_{sim}(\mathbf{x}) \in \mathcal{Y}$ is evaluated using a SPICE-level circuit simulation. We want to find an approximation

$$\hat{\mathbf{y}} = \mathbf{f}_m(\mathbf{x}) \in \mathcal{Y} : \min_{\mathbf{f}_m} E(\|\mathbf{y} - \hat{\mathbf{y}}\|^2), \tag{1.1}$$

such that the function $\mathbf{f}_m$ is much cheaper to evaluate than $\mathbf{f}_{sim}$ in terms of computational cost. In this chapter, unless specifically mentioned, we will now consider only any one output y_i at a time, from the vector $\mathbf{y}$. This is for the sake of clarity of explanation, and we will drop the subscript i and use only y. Then, for the output y, we can write (1.1) as

$$\hat{y} = f_m(\mathbf{x}) \in \mathbb{R} : \min_{f_m} E(\|y - \hat{y}\|^2). \tag{1.2}$$

If we use the L_2 norm in (1.2), we achieve the least squared error fit. To obtain this model in practice, some n sample points $\{\mathbf{x}_i, y_i = f_{sim}(\mathbf{x})\}_{i=1}^n$ are generated using SPICE simulations and the following optimization problem is solved.

$$\min_{f_m} \sum_{i=1}^n |y_i - \hat{y}_i|^2 \quad \text{where } \hat{y}_i = f_m(\mathbf{x}). \tag{1.3}$$

1.2.1 Linear Model

Linear models, such as the one used in [CYMSC85], model the response y as a linear function of the parameters $\mathbf{x}$. Hence, a linear model can be written as

$$\hat{y} = \mathbf{a}^T \mathbf{x} + c, \tag{1.4}$$

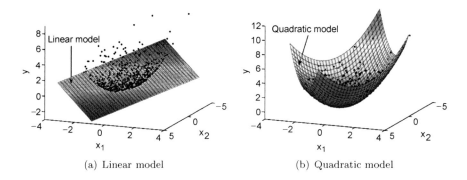

(a) Linear model (b) Quadratic model

Figure 1.1. A linear RSM cannot capture the quadratic behavior, while the quadratic RSM succeeds

where $\mathbf{a}$ is a vector of s unknown model parameters, $\mathbf{a} \in \mathbb{R}^s$ and c is an unknown real scalar. The total number of distinct, unknown model parameters is $n_p = s + 1 = O(s)$. Given $n \geq n_p$ training sample points, we can estimate $\mathbf{a}$ and c, using the least squares form of (1.3), as

$$[\mathbf{a}^T \ c]^T = [\mathbf{X} \ \mathbf{1}]^+ \mathbf{Y}, \quad \text{where } \mathbf{A}^+ = (\mathbf{A}^T \mathbf{A})^{-1} \mathbf{A}^T, \tag{1.5}$$

where $\mathbf{X}$ is an $n \times s$ matrix with the i-th row being the i-th input sample point $\mathbf{x}_i^T$ and $\mathbf{Y}$ is an n-vector with the i-th element being the i-th output sample point. $[\mathbf{X} \ \mathbf{1}]$ means $\mathbf{X}$ is augmented by a column of ones. Figure 1.1(a) shows an example of a linear model with $s = 2$: we can immediately see that the linear model cannot capture the nonlinear relationship in the data and the errors are very large.

1.2.2 Quadratic Model

Quadratic RSMs were proposed in [YKHT87][FD93] to model nonlinearities when the linear model fails. The quadratic model can be written as

$$\hat{y} = \mathbf{x}^T \mathbf{A} \mathbf{x} + \mathbf{b}^T \mathbf{x} + c, \tag{1.6}$$

where $\mathbf{A}$ is a symmetric $s \times s$ matrix of unknowns, $\mathbf{b}$ is a vector of s unknowns and c is an unknown scalar. The total number of distinct, unknown model parameters is $n_p = {}^{s+2}C_2 = (s+1)(s+2)/2 = O(s^2)$. Hence, the number of parameters grows quadratically with the number of dimensions. If we let $\mathbf{A}_i$ be the i-th row vector of $\mathbf{A}$ (written as a column vector), and the Kronecker product

$$\mathbf{x} \otimes \mathbf{x} = [x_1^2 \ x_1 x_2 \ \ldots \ x_1 x_s \ x_2 x_1 \ \ldots \ x_s^2]^T,$$

we can write (1.6) as

$$\hat{y} = \mathbf{a}_e^T \mathbf{x}_e, \quad \text{where } \mathbf{a}_e = [\mathbf{A}_1^T \ \ldots \ \mathbf{A}_s^T \ \mathbf{b}^T \ c]^T, \ \mathbf{x}_e = [(\mathbf{x} \otimes \mathbf{x})^T \ \mathbf{x}^T \ 1]^T, \tag{1.7}$$

which is similar in form to the linear model (1.4). Then, given $n \geq n_p$ training points, the least squared error estimate for the unknowns in (1.6) can be computed as

$$\mathbf{a}_e = \mathbf{X}_e^+ \mathbf{Y}, \tag{1.8}$$

where the i-th row of the matrix $\mathbf{X}_e$ is the vector $\mathbf{x}_e$ computed for the i-th input training point, and $\mathbf{X}_e^+$ is its pseudoinverse as in (1.5). In practice, the common (repeated) terms in $\mathbf{x} \otimes \mathbf{x}$ are combined, for example, $x_1 x_3$ and $x_3 x_1$. From Fig. 1.1(b) we can see that a quadratic model fits the data much better than a linear model. In this case the data was generated from a quadratic function of the two variables, and so we get a near-exact fit with a quadratic model. This full quadratic model can have a large number of unknowns and, hence, require a large number of training points n for proper fitting. Also, the computations in (1.8) can be very expensive for large s. This high fitting cost can be alleviated by using a *reduced-rank* quadratic model, like PROBE [LLPS05] reviewed next.

1.2.3 PROjection Based Extraction (PROBE): A Reduced-Rank Quadratic Model

A reduced-rank quadratic RSM was proposed by Li et al. in [LLPS05] to overcome the dimensionality problems of the full quadratic model. The matrix $\mathbf{A}$ in (1.6) is replaced by a low-rank approximation $\mathbf{A}_L$, given by

$$\mathbf{A}_L = \sum_{i=1}^{r} \lambda_i \mathbf{p}_i \mathbf{p}_i^T, \quad r < s, \tag{1.9}$$

where λ_i is the i-th dominant (largest) eigenvalue of $\mathbf{A}$ and $\mathbf{p}_i$ is the corresponding normalized eigenvector. It is known that this approximation minimizes the Frobenius-norm error for a given r: it is the optimal rank-r approximation [GL96]. Then the reduced quadratic model can be written as

$$\hat{y} = \sum_{i=1}^{r} \mathbf{x}^T \lambda_i \mathbf{p}_i \mathbf{p}_i^T \mathbf{x} + \mathbf{b}_i \mathbf{x} + c_i. \tag{1.10}$$

Thus, the model is a combination of r simpler quadratic models, where the i-th quadratic part varies along the projection along the i-th eigenvector $\mathbf{p}_i$. This is similar in flavor to the concept of projection pursuit that is used in a more powerful and flexible form for the SiLVR model proposed in this thesis.

Algorithm 1.1 The PROBE algorithm

Require: training sample points $\{\mathbf{x}_j, y_j\}_{j=1}^n$
1: **for** i = 1 to r **do**
2:     $g_i(\mathbf{x}) \leftarrow$ getRankOneQuadratic($\{\mathbf{x}_j, y_j\}_{j=1}^n$)
3:     **for all** sample points $\{\mathbf{x}_j, y_j\}$ **do**
4:         $y_j \leftarrow y_j - g_i(\mathbf{x}_j)$
5:     **end for**
6: **end for**
7: The rank-r model is $\hat{y} = \sum_{i=1}^r g_i(\mathbf{x})$

Algorithm 1.2 PROBE: `getRankOneQuadratic(`$\{\mathbf{x}_j, y_j\}_{j=1}^n$`)` function to extract the rank-1 estimate

Require: ϵ, a predefined tolerance
1: Randomly select $\mathbf{q}_0 \in \mathbb{R}^s$
2: $k = 0$, $\psi_0 = \infty$
3: **repeat**
4:     $k = k + 1$
5:     $\mathbf{q}_{k-1} = \mathbf{q}_{k-1} / \|\mathbf{q}_{k-1}\|_2$
6:     Solve (least squares error) $\min_{\mathbf{q}_k, \mathbf{b}_k, c} \psi_k$, where $\psi_k = \sum_{j=1}^n [y_j - (\mathbf{x}_j^T \mathbf{q}_k \mathbf{q}_{k-1}^T \mathbf{x}_j + \mathbf{b}_k \mathbf{x}_j + c_k)]^2$
7: **until** $|\psi_k - \psi_{k-1}| < \epsilon$
8: Return rank-1 estimate $g(\mathbf{x}) = \mathbf{x}^T \mathbf{q}_k \mathbf{q}_{k-1}^T \mathbf{x} + \mathbf{b}_k \mathbf{x} + c_k$

Since the matrix **A** is not already known, an implicit method that does not need it is used in [LLPS05] to estimate its eigenvectors. The overall algorithm is shown as Algorithm 1.1. Once the i-th component in (1.10) is extracted, the predicted $g_i(\mathbf{x})$ values for all the sample points are subtracted out, so that the $(i+1)$-th component fits the residual $y = y - g_i(\mathbf{x})$. Algorithm 1.2 extracts the i-th component using an implicit power iteration method, and constitutes the function `getRankOneQuadratic()` in Algorithm 1.1. The vector $\mathbf{q}_k \to \sqrt{\lambda_i}\mathbf{p}_i$ with $k \to \infty$ in Algorithm 1.2, for the i-th call to `getRankOneQuadratic()` in Algorithm 1.1. For a detailed explanation of the technique please refer to [LLPS05].

A rank-r quadratic model is effective in reducing the number of unknown model parameters and scales well with the number of dimensions s, if $r \ll s$: the number of model parameters is $n_p = 2r(s+1) = O(rs)$, which increases linearly with s. The authors of [LLPS05] show that r is very small for the performance metrics of some commonly seen circuits: even a rank-1 model can suffice. However, the model still suffers from a quadratic behavior assumption. We will now review some techniques that, in the general case, make no assumption regarding the modeled behavior, and then show how we can maintain much of this generality using the proposed SiLVR model.

1.3 Latent Variables and Ridge Functions

For the rest of this chapter let us assume that all the training sample points have been normalized – scaled and translated to mean 0 and variance 1 – in both the input and output spaces. This is for the sake of clear development of the following concepts, without any loss of generality.

1.3.1 Latent Variable Regression

With the assumption of normalized training points, the standard linear model for the s_y-vector of outputs $\mathbf{y}$ can be written as

$$\hat{\mathbf{y}} = \mathbf{A}\mathbf{x}, \tag{1.11}$$

where $\mathbf{A}$ is an $s_y \times s$ matrix of regression coefficients. Classically, latent variable regression (LVR) has been used to modify this linear model into a *reduced* linear model as

$$\hat{\mathbf{y}} = \mathbf{Z}\mathbf{W}_r\mathbf{x}. \tag{1.12}$$

Here $\mathbf{W}_r$ is an $r \times s$ matrix that projects the s-dimensional vector $\mathbf{x}$ to an r-dimensional space, where $r < s$, and $\mathbf{Z}$ is an $s_y \times r$-vector of regression coefficients over this reduced r-dimensional space. If we denote the i-th row of $\mathbf{W}_r$ by $\mathbf{w}_i$, then we can interpret $\mathbf{w}_i^T\mathbf{x}$ as the i-th coordinate in the reduced r-dimensional space. We will refer to $\mathbf{w}_i$ as the i-th *projection vector*, and the new variable $\mathbf{w}_i^T\mathbf{x}$ as the i-th *latent variable* t_i. Each coordinate w_{ij} of $\mathbf{w}_i$ will be referred to as the j-th *projection weight* of the i-th projection vector.

$$t_i = \mathbf{w}_i^T\mathbf{x}. \tag{1.13}$$

$\mathbf{W}_r$ is, then, the *projection matrix*.

The unknown parameters (the projection vectors $\mathbf{w}_i$ and the regression coefficients in $\mathbf{Z}$) can be chosen to satisfy a variety of criteria, each yielding a different LVR method (e.g., RRR, PLS, CCR) as shown in [BVM96]. The relevant method here is reduced rank regression (RRR), which solves the least squared error problem

$$\min_{\mathbf{W}_r, \mathbf{Z}} \|\mathbf{Y} - \mathbf{X}\mathbf{W}_r^T\mathbf{Z}^T\|^2, \tag{1.14}$$

where $\mathbf{X}, \mathbf{Y}$ are matrices of n sample points: each row is one sample point. From the discussion until now, the important idea to remember is that we are extracting the r statistically most important LVs ($\{t_1,\ldots,t_r\}$), such that the expected squared error is minimized, as in (1.14).

The problem of modeling nonlinear behavior, however, remains unsolved by these classical LVR techniques. Kernel-based methods try to address this issue by using the well-known "kernel trick": map the inputs ($\mathbf{x}$), using *fixed* nonlinear kernels ($f_K(\mathbf{x})$, e.g., a quadratic as in [FL06]), to a higher dimensional space, and then create a reduced linear model from this higher dimensional space to the output y [HTF01]. This has severe limitations: it increases the problem dimensionality before reducing it, and, more importantly, assumes a known nonlinear relationship between $\mathbf{x}$ and y. Baffi [BMM99] proposes adapting LVR to use a more flexible neural network [Rip96] formulation, but the model fitting is very slow (a two-step process that iterates between model fitting and LV estimation) and unreliable (due to weak convergence of this two-step iteration). Malthouse [MTM97] takes this further, but produces a very complex neural network model that can cause undesirable overfitting, especially for small training datasets, and has a large number of unknowns to fit. Also, both these methods solve a problem different from minimizing the least squared error as in (1.3). As we saw in Sect. 1.2.3, the PROBE method also uses a projection-based approach, but is restricted to a quadratic form.

The advantages of a flexible nonlinear LVR method are multiple and significant:

- It inherently reduces the dimensionality of the problem by extracting the LVs.

- The LVs are the "hidden variables" in the input space that impact the output in decreasing order of importance. Having this information can be of much use to the designer, as we shall see in the next few sections.

- The model would not be restricted to a small class of nonlinear behaviors.

All these features are very useful for addressing the problems mentioned in Sect. 1.1, and we will construct the SiLVR model to exploit all of them. First, though, we review the idea of projection pursuit, which bears close resemblance to LVR, and provides some theoretical foundation for the SiLVR model.

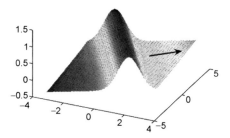

Figure 1.2. Example of a ridge function. The *arrow* indicates the projection vector

1.3.2 Ridge Functions and Projection Pursuit Regression

Projection pursuit regression (PPR) is a class of curve fitting algorithms, formally introduced first by Friedman and Stuetzle in [FS81] that approximate the output y as

$$\hat{y} = \sum_{i=1}^{r} g_i(\mathbf{w}_i^T \mathbf{x}), \qquad (1.15)$$

where, $\mathbf{w}_i$ is the i-th projection vector, similar to LVR, and $g_i : \mathbb{R} \to \mathbb{R}$ are unknown functions that might be parameterized functional forms (e.g., quadratic) or some nonparametric function, as in [FS81]. Hence, y is represented as the sum of nonlinear, univariate functions g_i, each varying along a different direction $\mathbf{w}_i$ in the input space. Each g_i function is called a *ridge function* [LS75] because for $s = 2$ it defines a 2-dimensional surface that is constant along one direction in the input space $\mathbb{R}^2$ (orthogonal to $\mathbf{w}_i$), leading to "ridges" in the topology. An example is shown in Fig. 1.2. In higher dimensions, a ridge function g_i is constant along the hyperplanes $\mathbf{w}_i^T \mathbf{x} = c$. Ridge functions have also been referred to as *plane waves* [VK61] historically, particularly in the field of partial differential equations [Joh55]. The representation in (1.15) is computed so as to minimize the modeling error as in (1.3). Given n training sample points, we can write this criterion as

$$\min_{r, \{g_i, \mathbf{w}_i\}_1^r} \sum_{j=1}^{n} \left\| y_j - \sum_{i=1}^{r} g_i(\mathbf{w}_i^T \mathbf{x}_j) \right\|^2. \qquad (1.16)$$

From (1.15), we can see the similarity to LVR, where we are also trying to extract r best directions to predict the output. In fact, a *nonlinear* version of LVR optimizing (1.3) will accomplish precisely the same thing as PPR.

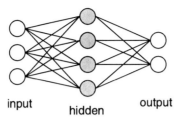

Figure 1.3. A feedforward neural network with one hidden layer: a 3-layer perceptron

The representation of (1.15) is also a general form of a feedforward neural network with one hidden layer. Artificial neural networks were introduced first by McCullough and Pitts in [MP43] to model the behavior of neurons in the nervous system. We will refer to them as simply neural networks. Since then, neural networks have been the focus of much theoretical and applied research [IM88][Fun89][Bar93][CS96][HSW89][Mha96] [HM94][FH97][NW90], and have been proposed in a large variety of forms [Rip96]. Here we refer to the simple feedforward form with one hidden layer of r nodes, which can be written mathematically as

$$\sum_{i=1}^{r} \alpha_i \sigma(\mathbf{w}_i^T \mathbf{x} + \beta_i), \quad \alpha_i, \beta_i \subset \mathbb{R}, \tag{1.17}$$

where $\sigma : \mathbb{R} \to \mathbb{R}$ is a fixed univariate function called the *activation function*. One such network is shown in Fig. 1.3. We will refer to such networks as 3-layer perceptrons (3LP) as per [IM88]: the first layer is just the layer of input nodes, layer two consists of the activation function nodes and layer three consists of the output node. From (1.17), we can immediately see the similarity with the PPR model of (1.15). Hence, a 3LP is a special case of a PPR model. We will revisit the 3LP when we develop the SiLVR model, where we use it in a somewhat different manner.

Before we proceed further, let us look at a couple of simple examples to clarify the concept of PPR. Consider the functions

$$y_1 = (x_1 + 2x_2)^3, \quad y_2 = x_1 x_2, \tag{1.18}$$

of which the second appears also in [DS84] and [Hub85]. We can represent the first function as

$$y_1 = t_1^3, \quad \text{where } t_1 = (1\ 2)\begin{pmatrix} x_1 \\ x_2 \end{pmatrix}. \tag{1.19}$$

In this case projection along only one direction $\mathbf{w}_1 = (1, 2)$ is enough to model the entire function exactly. This is because the function varies

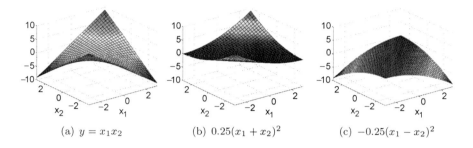

Figure 1.4. The function of (1.20) and its component ridge functions

only along that one direction. Hence, we have reduced the dimensionality of the input space to one. t_1 is the first LV, following the nomenclature from latent variable regression. On first glance, the second function could seem unfriendly to such linear projection-based decomposition. However, we can write y_2 as

$$y_2 = x_1 x_2 = 0.25(x_1 + x_2)^2 - 0.25(x_1 - x_2)^2, \qquad (1.20)$$

which is the sum of two univariate ridge functions (quadratics) along the directions $\mathbf{w}_1 = (1, 1)$ and $\mathbf{w}_2 = (1, -1)$, in the form of (1.15). The functions are shown in Fig. 1.4.

It is interesting to note that the Fourier series representation of a function,

$$f(\mathbf{x}) = \sum_{k=1}^{r} a_k e^{i \bar{\omega}_k^T \mathbf{x}}, \qquad (1.21)$$

is also a ridge function representation, where the projection vectors $\bar{\omega}_k^T$ are points in the s-dimensional Fourier domain. Note that the i in this equation is the imaginary unit, and not an index. Section 1.4 discusses a theorem from [DS84] that deals with representations similar to this.

Of course, for some unknown function f we would need to automatically extract the optimal projection directions and the corresponding ridge function. This "pursuit" of the optimal projections leads to the name projection pursuit. Before we discuss the algorithmic details of PPR, let us review some relevant results from approximation theory that establish a theoretical foundation for approximation using ridge functions. The reader who is more interested in the algorithmic considerations may skip forward to Sect. 1.5.

1.4 Approximation Using Ridge Functions: Density and Degree of Approximation

Before we can begin to develop algorithms for PPR, some more fundamental questions regarding ridge functions deserve attention. What can we approximate using ridge functions? How well can we approximate? To address these questions, let us first review some basic terminology from topology.

- $C(X)$ – For some space or set X, $C(X)$ is the set of all continuous functions defined on X, $f(\mathbf{x}|\mathbf{x} \in X)$.

- ***p*-norm** – Given some function f over some space X, we define the p-norm as

$$\|f\|_p = \begin{cases} \left(\int_X |f(\mathbf{x})| d\mathbf{x} \right)^{1/p}, & 0 < p < \infty \\ \sup_{\mathbf{x} \in X} |f(\mathbf{x})|, & p = \infty \end{cases}. \quad (1.22)$$

A norm taken over a set $D \subset X$ will involve integrating or taking the supremum, as relevant, over only D. The norm is then denoted by $\|f\|_{p,D}$, unless obvious.

- $L_p(X)$ – For some space or set X, $L_p(X)$ is the set of continuous functions defined over X that have a finite p-norm over X, equivalently

$$L_p(X) = \{f : \|f\|_{p,X} < \infty\}. \quad (1.23)$$

- **Compact set** – A compact set in Euclidean space $\mathbb{R}^s$ is any subset of $\mathbb{R}^s$ that is closed and bounded. A more general definition for any space is as follows. A set D is compact if for every collection of open sets $\mathcal{U} = \{U_i\}$ such that $D \subset \bigcup_i U_i$, there is a *finite* subset $\{U_{i_j} : j = 1, \ldots, m\} \subset \mathcal{U}$ such that $D \subset \bigcup_{j=1,\ldots,m} U_{i_j}$. For example, the *closed* unit ball $\{\mathbf{x} : \|\mathbf{x}\|_2 \leq 1, \mathbf{x} \in \mathbb{R}^s\}$ is compact, while the *open* unit ball $\{\mathbf{x} : \|\mathbf{x}\|_2 < 1, \mathbf{x} \in \mathbb{R}^s\}$ is not compact.

- **Dense set** – Let $V_1 \subset V_2$ be two subsets of some space V. Then, V_1 is dense in V_2 if for any $v \in V_2$ and any $\epsilon > 0$, there is a $u \in V_1$ such that $\|v - u\| < \epsilon$ under the p-norm specified or assumed without confusion. For example, if V_2 is $C[0,1]$ and V_1 is the space of all polynomials over $[a,b]$ then V_1 is dense in V_2 over $[a,b]$ under the ∞-norm, because every continuous function can be arbitrarily well approximated by polynomials, over some intervals $[a,b]$. This is the well-known Weierstrass approximation theorem [BBT97].

We now provide some answers to the questions posed at the beginning of this section, by reviewing relevant results from the approximation theory of ridge functions.

1.4.1 Density: What Can Ridge Functions Approximate?

THEOREM 1.1 (Diaconis and Shahshahani [DS84]). *Functions of the form $\sum \alpha_i e^{\mathbf{w}_i^T \mathbf{x}}$ with $\alpha_i \in \mathbb{R}$, $\mathbf{w}_i \in \mathbb{Z}_+$ are dense in $C[0,1]^s$ under the ∞-norm.*

This theorem says that any continuous function over the unit cube in s dimensions can be arbitrarily well approximated by ridge functions of the exponential form. Even though this theorem restricts itself to exponential ridge functions, it does prove that there exists a ridge function representation (linear combination of exponentials) for any continuous function over the unit cube. Note that the unit cube domain can be relaxed to any compact set D in s dimensions by including all required (continuous) transformations in the function to be approximated.

We will present a proof here, since it is simple enough for the non-mathematician to follow, while at the same time it provides some good insight and is an interesting read. The proof will require the following, very well-known, Stone–Weierstrass theorem, a generalization of the Weierstrass theorem. For a proof of the Stone–Weierstrass theorem please refer to standard textbooks on analysis, e.g., [BBT97]. Here we state a less general version of the theorem that suffices for our purposes.

THEOREM 1.2 (Stone–Weierstrass). *Let $D \subset \mathbb{R}^s$ be a compact set, and let V be a subspace of $C(D)$, the space of continuous functions on D, such that*

a) *V contains all constant functions,*

b) *$u, v \in V \Rightarrow uv \in V$, and*

c) *For every pair $\mathbf{x}, \mathbf{y} \in D$, $\mathbf{x} \neq \mathbf{y}$, $\exists v \in V$ such that $v(\mathbf{x}) \neq v(\mathbf{y})$.*

Then, V is dense in $C(D)$; i.e., $\forall v \in C(D)$, $\epsilon > 0$, $\exists u \in V$ such that $\|u - v\|_\infty$.

The Stone–Weierstrass theorem gives conditions such that V, the linear span of a given set of functions, is able to approximate any continuous function arbitrarily well. Using this, we can now prove Theorem 1.1.

PROOF OF THEOREM 1.1. Here V is the space of functions of the form $\sum \alpha_i e^{\mathbf{w}_i^T \mathbf{x}}$ with $\alpha_i \in \mathbb{R}$, $\mathbf{w}_i \in \mathbb{Z}_+$, and it satisfies the three conditions in the Stone–Weierstrass theorem, for the compact set $D = [0,1]^s$:

a) $\alpha_i e^{\mathbf{0}^T \mathbf{x}}$ are all the constant functions, where $\mathbf{0}$ is the vector of zeros.

b) $\sum_i \alpha_i e^{\mathbf{w}_i^T \mathbf{x}} \cdot \sum_j \beta_j e^{\mathbf{v}_j^T \mathbf{x}} = \sum_i \sum_j \alpha_i \beta_j e^{(\mathbf{w}_i + \mathbf{v}_j)^T \mathbf{x}}$, which lies in V.

c) For any pair $\mathbf{x}, \mathbf{y} \in [0,1]^p$ such that $\mathbf{x} \neq \mathbf{y}$, we must have $x_i \neq y_i$ for at least one $i \in 1, \ldots, s$. Then choose $\mathbf{w} = \mathbf{e}_i$, where the vector $\mathbf{e}_i$ is the unit vector along coordinate i. Then $e^{\mathbf{w}^T \mathbf{x}} = e^{x_i} \neq e^{\mathbf{w}^T \mathbf{y}} = e^{y_i}$.

Hence, the Stone–Weierstrass theorem applies and V is dense in $C[0,1]^s$.

More general results regarding the density of ridge functions have been developed by several authors, notably Vostrecov and Kreines [VK61], Sun and Cheney [SC92], and Lin and Pinkus [LP93]. Let $\mathcal{W} \subseteq \mathbb{R}^s$ be the set of possible projection vectors, and define

$$\mathcal{R}_\mathcal{W} = \text{span}\{g(\mathbf{w}^T \mathbf{x}) : \mathbf{w} \in \mathcal{W}, \ g \in C(\mathbb{R}), \ \mathbf{x} \in \mathbb{R}^s\} \tag{1.24}$$

as the linear span of all possible ridge functions using univariate continuous functions along directions defined by the vectors in $\mathcal{W}$. Now we state two results that specify conditions on $\mathcal{W}$ such that $\mathcal{R}_\mathcal{W}$ is dense in $C(\mathbb{R}^s)$.

THEOREM 1.3 (Vostrecov and Kreines [VK61]). $\mathcal{R}_\mathcal{W}$ *is dense on* $C(\mathbb{R}^s)$ *under the ∞-norm over compact subsets of $\mathbb{R}^s$ if and only if the only homogeneous polynomial of s variables that vanishes on $\mathcal{W}$ is the zero polynomial.*

A homogeneous polynomial is a polynomial whose terms all have the same degree. For example, $x_1^5 + x_1^2 x_2^3$ is a homogeneous polynomial of degree 5, while $x_1^5 + x_1^2$ is not. This theorem states that elements from $\mathcal{R}_\mathcal{W}$ can approximate any continuous function over any compact subset of $\mathbb{R}^s$ if and only if there is no nonzero homogeneous polynomial of s variables that has zeros at every point in $\mathcal{W}$. If we are allowed to choose any projection vector from $\mathbb{R}^s$; i.e., $\mathcal{W} = \mathbb{R}^s$, this is certainly true – the only homogeneous polynomial that is zero everywhere on $\mathbb{R}^s$ is the zero polynomial. Sun and Cheney state a similar, possibly simpler to visualize result:

THEOREM 1.4 (Sun and Cheney [SC92]). *Let $s \geq 2$ and let A_1, A_2, ..., A_s be subsets of $\mathbb{R}$. Put $\mathcal{W} = A_1 \times A_2 \times \cdots \times A_s$. $\mathcal{R}_\mathcal{W}$ is dense on*

$C(\mathbb{R}^s)$ under the ∞-norm over compact subsets of $\mathbb{R}^s$ if and only if at most one of the sets A_i is finite, and this finite set, if any, contains a nonzero element.

Once again, if $\mathcal{W} = \mathbb{R}^s$, this condition is obviously met. These two theorems state necessary and sufficient conditions for the same outcome. Hence, the conditions must be equivalent. In fact, it is easy to see that the condition in Theorem 1.4 is sufficient for the condition in Theorem 1.3. If all sets A_i are infinite, then no nonzero homogeneous polynomial of s variables can vanish everywhere on $\mathcal{W} = \prod_i A_i$, since it would need to have an infinite number of roots. Now, consider the case that one set A_i is finite, with at least one nonzero element. Any nonzero homogeneous polynomial h_s of s variables with no x_i term would be a homogeneous polynomial h_{s-1} of $s-1$ variables. This h_{s-1} would then not vanish over $\prod_{j \neq i} A_j$, by the same argument, and, so neither would h_s vanish over $\mathcal{W}$. The only case that remains is when h_s does contain a term with x_i. If h_s now vanished everywhere on $\mathcal{W}$, it would vanish also at all points with x_i equal to a nonzero value from A_i. Replacing this value for x_i in h_s again gives us a homogeneous polynomial h_{s-1} in $s-1$ variables. Hence, by the same argument as before any nonzero h_s cannot vanish over $\mathcal{W}$.

These theorems answer the first question we asked at the beginning of this section: what can we approximate using ridge functions? The answer is essentially, any nonlinear function we are likely to encounter in practice. Now, we look at some results that try to answer the second question: how well can we approximate?

1.4.2 Degree of Approximation: How Good Are Ridge Functions?

One way to address the question, "how well can ridge functions approximate?", is to study the convergence of approximation using ridge functions – how does the error decrease as we increase the number of ridge functions in the model, or in other words, the model complexity? This is a difficult question for general ridge functions and there are some partial results here, notably [Pet98][Mai99][BN01][Mha92][Bar93]. Many of these results exploit constraints on the ridge functions to show the convergence behavior of the approximation.

From these, we state a general result, by Maiorov. Let $B^s = \{\mathbf{x} \in \mathbb{R}^s : \|\mathbf{x}\|_2 \leq 1\}$ be the closed unit ball. Let $W_2^{k,s}$ be a Sobolev class [Ada75] of functions from $L_2(B^s)$. This is the class of functions $f \in L_2(B^s)$, for which all partial derivatives $\nabla_{\mathbf{x}}^{\mathbf{v}} f$ of order smaller than or equal to k ($\sum_{i=1}^s v_i \leq k$, where $\mathbf{v} = \{v_1, \ldots, v_s\}$), satisfy $\|\nabla_{\mathbf{x}}^{\mathbf{v}} f\|_{2, B^s} \leq 1$. These partial derivatives are taken in the weak sense [Ada75]. Define

$$\mathcal{R}_r = \left\{ \sum_{i=1}^{j} g_i(\mathbf{w}_i^T \mathbf{x}) : j \leq r, \ \mathbf{w}_i \in \mathbb{R}, \ g \in L(\mathbb{R}) \right\}, \quad (1.25)$$

where by $L(\mathbb{R})$, we mean the space of all functions integrable on any compact subset of $\mathbb{R}$, or equivalently, there is some compact set $D \in \mathbb{R}$ such that $g \in L_1(D)$. For any two sets of functions, U, V, define the distance of U from V as

$$dist(U, V) = \sup_{u \in U} \inf_{v \in V} \|u - v\|_2. \quad (1.26)$$

In words, for any given $u \in U$ find the distance to the closest approximation v from V using the 2-norm. Then find the maximum of this distance over all possible $u \in U$. Hence, if U is the target set of functions to be approximated and V is the set of possible approximations, this metric computes the maximum error, using the best possible approximations from V. It then follows that $dist(W_2^{k,s}, \mathcal{R}_r)$ is the maximum error while approximating functions in $W_2^{k,s}$ using best fitting approximations from $\mathcal{R}_r$. Now, we are equipped to state the following.

THEOREM 1.5 (Maiorov [Mai99]). *For $k > 0$, $s \geq 2$, the following asymptotic relation holds.*

$$dist(W_2^{k,s}, \mathcal{R}_r) = \Theta(r^{-k/(s-1)}). \quad (1.27)$$

Here, Θ is the tight bound notation [CLR01]. Hence, the maximum approximation error using r ridge functions decreases as $r^{-1/(s-1)}$, for a class of functions that satisfy a given smoothness criterion ($f \in W_2^{k,s}$).

All the results stated in this section provide us with some confidence that a ridge function-based approximation is theoretically feasible. [Lig92] surveys some methods of constructing the approximation $\hat{y}$ if the original function y is known. However, all these results deal with functions and not with finite sample sets. In a practical response surface model generation scenario we would not know anything about the behavior of the function we are trying to approximate, but we would have a finite set of points from which we have to estimate the "best" projection vectors and functions in the RSM in (1.16). The projection pursuit regression technique strives to accomplish precisely this with a statistical perspective. The next section reviews the original projection pursuit algorithm and some relevant convergence results.

Algorithm 1.3 The projection pursuit regression algorithm of Friedman and Stuetzle [FS81]

Require: normalized training samples $\{\mathbf{x}_j, y_j\}_{j=1}^n$
1: $e_j \leftarrow y_j$, $j = 1, \ldots, n$ and $r = 0$
2: find $\mathbf{w}_{r+1}$ to maximize the fraction of variance explained by g_{r+1}:

$$I \leftarrow \max_{\mathbf{w}_{r+1} \in \mathcal{S}^{s-1}} 1 - \frac{\sum_{j=1}^n (e_j - g_{r+1}(\mathbf{w}_{r+1}^T \mathbf{x}_j))^2}{\sum_{j=1}^n e_j^2}. \quad (1.29)$$

g_{r+1} is the smooth along the direction $\mathbf{w}_{r+1}$. Rosenbrock's method [Ros60] was used for the search
3: **if** $I < \epsilon$ **then**
4:     return $\{g_i, \mathbf{w}_i\}_{i=1}^r$
5: **else**
6:     $e_j \leftarrow e_j - g_{r+1}(\mathbf{w}_{r+1}^T \mathbf{x}_j)$, $j = 1, \ldots, n$
7:     $r \leftarrow r + 1$
8:     go to step 3
9: **end if**

1.5 Projection Pursuit Regression

The PPR algorithm, as proposed by Friedman and Stuetzle [FS81], takes a nonparametric approach to solve for the functions g_i and projection vectors $\mathbf{w}_i$ in (1.16). Each g_i is approximated using a smoothing over the training data. Let $\{t_j, y_j\}_{j=1}^n$ be our training data projected along some projection vector $\mathbf{w}$. In general, a smoothing-based estimate uses some sort of local averaging:

$$g(t) = \text{AVE}_{t_j \in [t-h, t+h]}(y_j). \quad (1.28)$$

Here AVE can denote the mean, median, any weighted mean, or any other ways of averaging (e.g., nonparametric estimators in [Pra83]). The parameter h defines the *bandwidth* or the *smoothing window*. We call the function g a *smooth*. Specific details of the smoothing method used by Friedman and Stuetzle can be found in [FS81][Fri84]. Their overall PPR algorithm is shown as Algorithm 1.3. We remind the reader that all training data has been normalized to mean 0 and variance 1, and denote the surface of the unit sphere in $\mathbb{R}^s$ as $\mathcal{S}^{s-1}$. Hence, $\mathcal{S}^{s-1}$ is the set of all s-vectors of magnitude 1.

We can see that the algorithm is iterative. At each iteration, it tries to extract the best direction $\mathbf{w}_{r+1}$ and the corresponding ridge function g_{r+1} so as to best approximate the residue values $\{e_j\}$ at that iteration. We can clearly see the similarity with latent variable regression. The i-th latent variable in this case is the displacement along the i-th projection vector $t_i = \mathbf{w}_i^T \mathbf{x}$. This iterative approach simplifies the problem of extracting all the required projections and ridge functions, by handling only one component at a time. This has the advantage of scoping down the problem to a one-dimensional curve fitting problem, from a very difficult high-dimensional curve fitting problem. Furthermore, since each component is extracted to maximally model the residue at that iteration, the latent variable associated with the i-th projection vector can be interpreted as the i-th most important variable for explaining the output behavior. This can be very useful for extracting some deep insight into the behavior of a circuit, when PPR is used for RSM building. We will revisit this observation and elaborate further on it when we explain the SiLVR model.

1.5.1 Smoothing and the Bias–Variance Tradeoff

There is a subtle, but critical, observation we will make here regarding the ridge function that is extracted in any one iteration. This is best introduced using an illustration: we refer back to our example from (1.20), and reproduce it here in a slightly different form for the reader's convenience:

$$y_2 = x_1 x_2 = 0.25(\{1,1\} \cdot \mathbf{x})^2 - 0.25(\{1,-1\} \cdot \mathbf{x})^2. \quad (1.30)$$

Suppose the ridge function g_1 was unconstrained with regard to any smoothness requirement and was free to take up any shape. Then, given n training points, a perfect, zero-error interpolation could be performed along *any* direction $\mathbf{w}_1$. Figure 1.5 illustrates this. From (1.30) we know that $\mathbf{w}_1 = \{1,1\}$ or $\{1,-1\}$ are two good candidates for the first projection vector. In fact, any $\{a,b\}$ such that $ab \neq 0$ is a good candidate because we can write

$$x_1 x_2 = (4ab)^{-1}[(ax_1 + bx_2)^2 - (ax_1 - bx_2)^2]. \quad (1.31)$$

Therefore, $\{1,0\}$ is a bad projection vector. Figure 1.5 shows 100 training points as (blue) dots, projected along the projection vectors $\{1,0\}$ (Fig. 1.5(a)) and $\{1,1\}$ (Fig. 1.5(b)). With unrestricted g_1 we can find perfect interpolations along both directions, shown as solid lines joining the projected training points. In both cases, the metric I in step 3 of Algorithm 1.3 is maximized to 1 and the algorithm has no way of determining which is the better direction. In fact, with such a flexible class of functions for g_1, all directions will have $I = 1$. Also, once the first ridge function is extracted, the algorithm will stop because all the variance in the training data will have been explained and I would be 0 for the

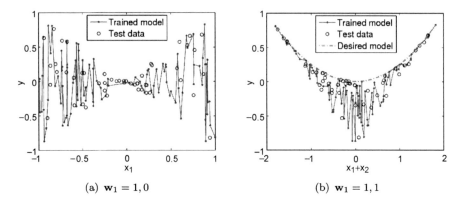

Figure 1.5. Overfitting of training data along two different projection vectors ($\mathbf{w}_1$) for $y = x_1 x_2$

second iteration, resulting in a final model with only one component ridge function model. The solid lines shown in Figs. 1.5(a) and 1.5(b) are, in fact, the final models. However, choosing the wrong projection vector $\mathbf{w}_1 = \{1, 0\}$ in Fig. 1.5(a) results in large errors on unseen test data, shown as black circles. This is, of course, as expected because the direction of projection is incorrect in the first place. However, even with the correct projection in Fig. 1.5(b), we get large errors on unseen test data.

The problem here is the unrestricted flexibility in the function g_1. A more desired g_1 along $\mathbf{w}_1 = \{1, 1\}$ is actually a very smooth function in this case, shown as a (red) dash-dot line in Fig. 1.5(b). This is the first term in expansion in (1.30). Note that this ridge function has large errors on the training data and does not try to exactly fit the training points along the projection. However, it lets the algorithm perform a second iteration, in which the second projection vector $\{1, -1\}$ is chosen and the second ridge function in (1.30) is extracted, giving us a near-exact two-component ridge function model. Such a class of smooth univariate functions will have a larger error along the incorrect direction of Fig. 1.5(a) and the algorithm will easily reject it. This illustrates the classic bias-variance tradeoff in statistical learning [HTF01]. If we minimize the bias in our estimated model by exactly fitting the training data, we will get a completely different approximation for a different set of training points, resulting in high variance. This choice also results in large errors on unseen points. If we minimize the variance, by estimating nearly the same model for different sets of training data, we need to reconcile with a larger training error. In the extreme version of this choice, any training sample will result in the same estimate of the model, meaning that we are not even using any information from the training data. Such extremes will also result in large errors on unseen data. Hence, we must find a balance such that we keep the error low on both the training data and on unseen test data. This is the classic problem of *generalization*.

This issue is particularly critical for the case of PPR. When we project the training data onto a single direction $\mathbf{w}_i$, there can be a lot of noise or variation in the output values because of smooth dependence on other directions orthogonal to $\mathbf{w}_i$, as in Fig. 1.5(b) for $\mathbf{w}_1$. If the function g_i is allowed too much flexibility, it will undesirably *overfit* the training data by fitting this orthogonal contribution to the behavior of the function. Hence, it is critical that any PPR algorithm employ some technique to avoid overfitting and improve the generalizability of the model. Friedman and Stuetzle used variable bandwidth smoothing to achieve this: the parameter h in (1.28) is adaptively changed to be larger in those parts of the projected input space where the function variation is estimated to be high, since this high variation is probably because of higher dependence on orthogonal directions in that region. Minimizing overfitting will a prime objective when we develop the proposed SiLVR model.

1.5.2 Convergence of Projection Pursuit Regression

PPR was proposed in [FS81] relying on intuitive arguments regarding why it should work and its advantages, as mentioned in the beginning of this section (Sect. 1.5). Unfortunately, the theoretical results developed for approximation using ridge functions (Sect. 1.4) do not directly apply to PPR because of at least two reasons. First, PPR uses a finite set of training points and does not have knowledge of the original function to be modeled. Second, PPR extracts each projection iteratively. Hence, it cannot rely on exact interpolation techniques, and must use statistical estimation. This was discussed in the context of the bias-variance tradeoff and smoothing in Sect. 1.5.1. Also, this iterative scheme is a "greedy" approach, where at every step only the next best decision is taken – to select the next best projection and ridge function. The best decision at any given iteration might not be the best decision in the global sense. It might be better sometimes to not choose the ridge function that seems to be the best for the current iteration. In fact, later in this section, we will show an example where the choice made by PPR does not match the best choice suggested by analysis. Given this greedy nature, does the algorithm still converge to a good solution (to an accurate RSM)? Researchers in statistics have recognized these issues and questions, and there are some theoretical results showing convergence of PPR under different conditions [Hub85][DJRS85][Jon87][Hal89]. In this section we review some of these results.

Any set of training points will be drawn from some underlying probability distribution defined over the sampling space $\mathcal{X}$. We denote this distribution by P, and the probability density is denoted by p. This scenario is reasonable for our applications, since any statistical parameter (e.g., V_t) or design variable will follow some probability distribution (e.g., normal distribution) or lie uniformly in some bounded range. A bounded domain $D \in \mathbb{R}^s$ can be represented as a uniform distribution P that is nonzero for subsets in D and zero for subsets outside D. Any expectation computation will then be performed over the relevant probability distribution, unless differently specified. For example, the expectation (mean) of a circuit performance $y = f(\mathbf{x})$ will be computed as

$$E(y) = \int_P f(\mathbf{x})dP = \int_{\mathcal{X}} f(\mathbf{x})p(\mathbf{x})d\mathbf{x}. \quad (1.32)$$

In general terms, P is the probability measure over the sample space $\mathcal{X}$ [Lo77]. For our circuit applications $\mathcal{X}$ is typically $\mathbb{R}^s$.

In practice, the PPR algorithm has to deal with at least three nonidealities:

1) No exact knowledge of the original function $y = f(\mathbf{x})$ – we have only a finite number of training points n.

2) Imperfect approximation technique for estimating the best univariate function g along any direction $\mathbf{w}$.

3) Imperfect search algorithm to search for the best $\mathbf{w}$ in any iteration.

To the best of our knowledge, there is no theoretical result establishing the convergence properties of PPR in the most general case allowing for all these nonidealities. However, there are results that make ideality assumptions for one or more of the three points mentioned above, but still provide insight into the general working of PPR.

Let us assume that we have a perfect version of PPR, free of the three nonidealities mentioned above. Then we ask the question,

> What are the *best* projection vector $\mathbf{w}$
> and the *best* univariate function g?

By *best* we mean the pair $(\mathbf{w}, g)$ that gives the best approximation; that is, minimizes the mean squared error. If we are in the i-th iteration, then we can define the residue e_{i-1} as

$$e_{i-1}(\mathbf{x}) = f(\mathbf{x}) - \sum_{j=1}^{i-1} g_j(\mathbf{w}^T \mathbf{x}). \quad (1.33)$$

Following (1.2), the best $(\mathbf{w}_i, g_i)$ will satisfy

$$(\mathbf{w}_i, g_i) = \arg\min_{\mathbf{w},g} E[(e_{i-1}(\mathbf{x}) - g(\mathbf{w}^T\mathbf{x}))^2]. \qquad (1.34)$$

Let us first assume some candidate $\mathbf{w}$, and ask,

> For any given projection vector $\mathbf{w}$,
> what is the best univariate function g?

From (1.34), we know that the best g_i will minimize the error in approximating the residue

$$g_i = \arg\min_{g} E[(e_{i-1}(\mathbf{x}) - g(\mathbf{w}^T\mathbf{x}))^2]. \qquad (1.35)$$

For every g, since $g(\mathbf{w}^T\mathbf{x})$ is constant $(= g(t))$ for all $\mathbf{w}^T\mathbf{x} = t$, we can write this criterion as follows. The best g_i will minimize the error in approximating the residue projected *along* $\mathbf{w}$:

$$g_i(t) = \arg\min_{g_t} E[(e_{i-1}(\mathbf{x}) - g_t)^2 | \mathbf{w}^T\mathbf{x} = t], \quad \forall t, \qquad (1.36)$$

where g_t is some scalar value. For any displacement t along the direction $\mathbf{w}$, we expect to see a distribution of values for the residue e_{i-1}, since multiple $\mathbf{x}$ will map to the same t. The best value of the new ridge function at t, $g_i(t)$, minimizes the mean squared error between the residue and g_i at t. The expectation here is taken over the marginal distribution of $\mathbf{x}$ in the hyperplane $\mathbf{w}^T\mathbf{x} = t$, which is a hyperplane normal to $\mathbf{w}$. This same criterion is applied for all t to obtain the complete function $g_i(t)$ for all values of t. Then, for any t, we can write

$$E[(e_{i-1}(\mathbf{x}) - g_t)^2 | \mathbf{w}^T\mathbf{x} = t] = E[e_{i-1}^2(\mathbf{x}) - 2e_{i-1}(\mathbf{x})g_t + g_t^2 | \mathbf{w}^T\mathbf{x} = t]$$

$$= E[e_{i-1}^2(\mathbf{x}) | \mathbf{w}^T\mathbf{x} = t]$$

$$- 2g_t E[e_{i-1}(\mathbf{x}) | \mathbf{w}^T\mathbf{x} = t] + g_t^2 \qquad (1.37)$$

since g_t is a constant for a given t. Then the optimal $g_i(t)$ for a given t is

$$g_i(t) = g_t : \frac{d}{dg} E[(e_{i-1}(\mathbf{x}) - g_t)^2 | \mathbf{w}^T\mathbf{x} = t] = 0$$

$$\Rightarrow \quad -2E(e_{i-1}(\mathbf{x}) | \mathbf{w}^T\mathbf{x} = t) + 2g_i(t) = 0$$

$$\Rightarrow \quad g_i(t) = E(e_{i-1}(\mathbf{x}) | \mathbf{w}^T\mathbf{x} = t). \qquad (1.38)$$

Thus, the best value of $g_i(t)$ is the expectation of the residual $e_{i-1}(\mathbf{x})$. We have, thus, proved the following theorem that appears in [Hub85]:

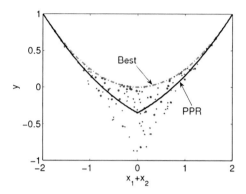

Figure 1.6. Optimal ridge functions from analysis (*red dash-dot*) and PPR (*black solid*) can differ. This example is for $y = x_1 x_2$, along the projection vector $\{1,1\}$

THEOREM 1.6. *For any given projection vector* $\mathbf{w}$, *the best function* $g_i(t)$ *defined by* (1.35), *is given by*

$$g_i(t) = E(e_{i-1}(\mathbf{x}) | \mathbf{w}^T \mathbf{x} = t). \tag{1.39}$$

This is an interesting result. The solution from this result can be quite different from what standard approximation theory would suggest. This is easily illustrated with our friendly example from (1.20) that is reproduced here for convenience:

$$y = x_1 x_2 = 0.25(\{1,1\} \cdot \mathbf{x})^2 - 0.25(\{1,-1\} \cdot \mathbf{x})^2. \tag{1.40}$$

Say we are considering one of the optimal directions $\mathbf{w}_1 = \{1,1\}$. To achieve an exact approximation, as per (1.40), the best $g_1(t)$ is

$$g_1(t) = 0.25 t^2, \tag{1.41}$$

which is just the first term on the right hand side of (1.40) mapped on to the latent variable t. This function is shown as the (red) dash-dot line in Fig. 1.6, and also previously in Fig. 1.5(b). It is indicated by "Best". However, the best $g_1(t)$ for PPR, as per Theorem 1.6, is the expectation of y taken over the hyperplane $\mathbf{w}_1^T \mathbf{x} = t$. Assuming that $\mathbf{x} \in [-1, 1]^2$, we can analytically compute this best g_1 function. This best g_1 is shown as the (black) solid line in Fig. 1.5(b) and is indicated by "PPR". We can clearly see that the two ridge functions are different. This difference is a result of PPR performing a greedy search by looking at only one projection at a time unlike the analysis in (1.40) which looks at the function as a whole over all the dimensions.

Given this optimal choice of g_i, we now ask,

> What then is the best projection vector $\mathbf{w}_i$ that will satisfy (1.34)?

From (1.34) and (1.39), we know that such a $\mathbf{w}_i$ must satisfy

$$\mathbf{w}_i = \arg\min_{\mathbf{w}} E[(e_{i-1}(\mathbf{x}) - g_i(\mathbf{w}_i^T\mathbf{x}))^2],$$
$$\text{where } g_i(t) = E(e_{i-1}(\mathbf{x})|\mathbf{w}^T\mathbf{x} = t). \quad (1.42)$$

Expanding the first expectation we get

$$E[(e_{i-1}(\mathbf{x}) - g_i(\mathbf{w}_i^T\mathbf{x}))^2] = E[e_{i-1}^2(\mathbf{x})] - 2E[e_{i-1}(\mathbf{x})g_i(\mathbf{w}_i^T\mathbf{x})]$$
$$+ E[g_i^2(\mathbf{w}^T\mathbf{x})]. \quad (1.43)$$

Since $g_i(\mathbf{w}_i^T\mathbf{x})$ is a constant for all $\mathbf{w}_i^T\mathbf{x}$ equal to some constant t (it is a ridge function along $\mathbf{w}_i$), we have

$$E[g_i^2(\mathbf{w}^T\mathbf{x})] = E[g_i^2(t)]. \quad (1.44)$$

Let us now expand out the second expectation term on the right hand side of (1.42) as

$$E[e_{i-1}(\mathbf{x})g_i(\mathbf{w}_i^T\mathbf{x})] = \int_{\mathcal{X}} e_{i-1}(\mathbf{x})g_i(\mathbf{w}_i^T\mathbf{x})p(\mathbf{x})d\mathbf{x}. \quad (1.45)$$

Let us denote the marginal probability density of any t along $\mathbf{w}_i$ as $p_{\mathbf{w}_i}(t)$. Also let $\mathcal{X}_{\mathbf{w}_i}$ denote the range of $t = \mathbf{w}_i^T\mathbf{x}$ for $\mathbf{x} \in \mathcal{X}$. For our circuit applications, typically $\mathcal{X} = \mathbb{R}^s$ so that $\mathcal{X}_{\mathbf{w}_i} = \mathbb{R}$. Given these definitions, we can rewrite (1.45) as

$$E[e_{i-1}(\mathbf{x})g_i(\mathbf{w}_i^T\mathbf{x})]$$
$$= \int_{t \in \mathcal{X}_{\mathbf{w}_i}} \left[\int_{\mathbf{x}:\mathbf{w}_i^T\mathbf{x}=t} e_{i-1}(\mathbf{x})g_i(\mathbf{w}_i^T\mathbf{x})p(\mathbf{x}|\mathbf{w}_i^T\mathbf{x}=t)d\mathbf{x}\right] p_{\mathbf{w}_i}(t)dt. \quad (1.46)$$

Since $g_i(\mathbf{w}_i^T\mathbf{x} = t)$ is a constant for a given t, we can take it out of the inner integral, giving

$$E[e_{i-1}(\mathbf{x})g_i(\mathbf{w}_i^T\mathbf{x})]$$
$$= \int_{t \in \mathcal{X}_{\mathbf{w}_i}} g_i(t) \left[\int_{\mathbf{x}:\mathbf{w}_i^T\mathbf{x}=t} e_{i-1}(\mathbf{x})p(\mathbf{x}|\mathbf{w}_i^T\mathbf{x}=t)d\mathbf{x}\right] p_{\mathbf{w}_i}(t)dt. \quad (1.47)$$

Now, the inner integral is nothing but $E(e_{i-1}(\mathbf{x})|\mathbf{w}^T\mathbf{x} = t)$, which is the same as $g_i(t)$ according to Theorem 1.6. Hence, we get

$$E[e_{i-1}(\mathbf{x})g_i(\mathbf{w}_i^T\mathbf{x})] = \int_{t \in \mathcal{X}_{\mathbf{w}_i}} g_i^2(t)p_{\mathbf{w}_i}(t)dt = E[g_i^2(t)]. \quad (1.48)$$

Substituting this and (1.44) in (1.42), we get

$$E[(e_{i-1}(\mathbf{x}) - g_i(\mathbf{w}_i^T\mathbf{x}))^2] = E[e_{i-1}^2(\mathbf{x})] - E[g_i^2(t)],$$
where $t = \mathbf{w}_i^T\mathbf{x}$. \hfill (1.49)

Hence, we have proved the following, which appears in [Hub85] without a proof:

THEOREM 1.7. *The optimal $\mathbf{w}_i$ of (1.34) is the one that maximizes the variance of the function g_i, where g_i is chosen as in Theorem 1.6.*

One would expect any approximation of a function f to maximally explain the variance of f, and this result shows that PPR tries to achieve precisely this.

In Sect. 1.4.1 we saw that the a ridge function approximation, like in (1.86), converges to the approximated function, but does the "greedy" and statistical PPR method converge? Jones addresses this question in [Jon87] and proves strong convergence of PPR, as stated by the following theorem. Here, we assume ideality for conditions 1) and 2) – we have infinite number of points to exactly compute expectations, and we can compute the exact best functions g_i along any given direction, respectively. However, we do allow for error in estimating the optimal projection vector.

THEOREM 1.8 (Jones [Jon87]). *Let $f(\mathbf{x}) \in L_2(P)$, where P is the probability measure (distribution) for $\mathbf{x} \in \mathbb{R}^s$. Let PPR choose any possibly sub-optimal $\mathbf{w}_r$ such that $E[g_r(\mathbf{w}_r^T\mathbf{x})^2] > \rho \cdot \sup_{\|\mathbf{b}\|_2=1, \mathbf{b}\in\mathbb{R}^s} E(g_r(\mathbf{b}^T\mathbf{x})^2]$ for some fixed ρ, $0 < \rho < 1$. Then, $e_r(\mathbf{x}) \to 0$, as $r \to \infty$.*

Hall, in [Hal89], proves a convergence result for a scenario closer to practical PPR, accounting for many nonidealities. The only ideality assumption about the algorithm is that the search for the optimal projection vector is perfect, within the constraint of a finite number of training points n. This means that the a sub-optimal projection may seem optimal because of the incomplete information from finite number of points, but the search algorithm will find this seemingly optimal projection. Also, the results in the paper are for the classical PPR technique [FS81] that employs some sort of smoothing (1.28) using a kernel function with window or bandwidth h, to estimate the function g along some direction $\mathbf{w}$. If K is the kernel function used and the training data set is

$\{\mathbf{x}_j, y_j\}_{j=1}^n$, then the estimate is given as

$$\hat{g}(t) = \frac{\sum_{j=1}^n y_j K[(t - \mathbf{w}^T \mathbf{x}_j)/h]}{\sum_{j=1}^N K[(t - \mathbf{w}^T \mathbf{x}_j)/h]}. \tag{1.50}$$

Hence, there is one kernel instance centered at the projection of every training point onto the vector $\mathbf{w}$ and the function value at any location along $\mathbf{w}$ is the weighted sum of contributions by each of these n kernels, the weights being the y_j output values associated with each kernel center. The bandwidth h determines the range of influence of each kernel center. Higher values of h lead to smoother estimates resulting in low variance error, but increase the bias error if h is too large. The denominator performs the appropriate normalization. This is the estimate used for prediction. A slightly different form of (1.50) is used for driving the search for the optimal projection vector. Please refer to [Hal89] for details. The kernel is taken to satisfy the condition

$$\int_{-\infty}^{\infty} t^i K(t) dt = \begin{cases} 1, & i = 0 \\ 0, & 1 \leq i \leq k-1 \end{cases}, \tag{1.51}$$

and the first $k+1$ directional derivatives of $p(\mathbf{x})$ and $f(\mathbf{x})$ exist and are continuous in $\mathbb{R}$. Under some more loose conditions on $p(\mathbf{x})$ and K, the following holds.

THEOREM 1.9 (Hall [Hal89]). *Let $\mathbf{w}$ and g be the optimal projection vector and ridge function for any PPR iteration, and $\hat{\mathbf{w}}$ and $\hat{g}$ be the sub-optimal estimates resulting from n training points and the imperfect kernel-based approximation of (1.50). Then, the error between $\hat{g}(\hat{\mathbf{w}}^T \mathbf{x})$ and $g(\mathbf{w}^T \mathbf{x})$ decreases as $O(n^{-k/(2k+1)})$ for appropriately chosen h.*

Implications of this result are discussed in [Hal89] and are not immediately relevant here. However, it shows that convergence can be achieved even with significant nonidealities in the PPR algorithm, nonidealities that are unavoidable in any practical implementation. We are now well-equipped to develop the proposed SiLVR model with some confidence.

1.6 SiLVR

In this section we describe the SiLVR model and its features in detail.

1.6.1 The Model

SiLVR implements PPR, but uses building blocks and training algorithms that are different from the classical PPR method of [FS81]. The

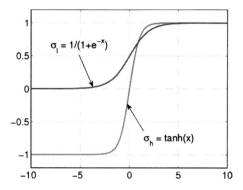

Figure 1.7.   Examples of sigmoidal functions

SiLVR model can be represented mathematically as a standard PPR model:

$$\hat{y} = \sum_{i=1}^{r} g_i(\mathbf{w}_i^T \mathbf{x}). \qquad (1.52)$$

However, the functions g_i are not purely nonparametric. In this case we use a linear combination of *sigmoidal* functions ($\sigma(\cdot)$) to represent g_i, as

$$g_i(t) = \sum_{j=1}^{q} a_{ij}\sigma(b_{ij}t + c_{ij}) : a_{ij}, b_{ij}, c_{ij} \in \mathbb{R}, \ i \in \{1, 2, \ldots\}, \ j \in \{1, \ldots, q\}, \qquad (1.53)$$

where q is the number of sigmoids used for the approximation of one ridge function. The complete model can then be written as

$$\hat{y} = f_{\text{SiLVR}}(\mathbf{x}) = \sum_{i=1}^{r}\sum_{j=1}^{q} a_{ij}\sigma(b_{ij}\mathbf{w}_i^T \mathbf{x} + c_{ij}). \qquad (1.54)$$

A sigmoidal function or sigmoid is typically defined as a continuous, monotonic function $\sigma(t)$ such that $\lim_{t\to\infty} \sigma(t) = 1$ and $\lim_{t\to-\infty} \sigma(t) = 0$. Any such function taken through scaling and translation is also a sigmoid. Standard examples of sigmoidal functions are the logistic function

$$\sigma_l(t) = \frac{1}{1 + e^{-t}}, \qquad (1.55)$$

and the hyperbolic tangent function

$$\sigma_h(t) = \tanh(t) = \frac{e^{2t} - 1}{e^{2t} + 1}. \qquad (1.56)$$

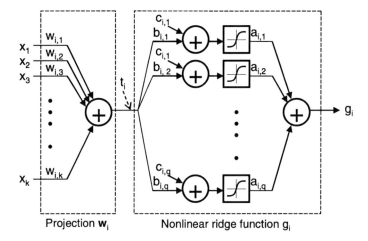

Figure 1.8. The network corresponding to the i-th component in the SiLVR model (1.54)

Both are shown in Fig. 1.7. In fact, these two sigmoids are equivalent in terms of their nonlinear approximation power using linear combinations because

$$\sigma_h(t) = 2\sigma_l(2t) - 1. \tag{1.57}$$

These functions have a very desirable property that their derivative is easily computed:

$$\frac{d\sigma_h}{dt} = 1 - \sigma_h^2. \tag{1.58}$$

This is useful for efficient training, which typically involves gradient computations within some optimizer. Section 1.6.4.2 shows how we exploit this property for SiLVR. Also, using a parametric model with few basis functions is much more efficient computationally than using data-dependent non-parametric methods like the PPR method of Friedman and Stuetzle. We will use the tanh function and refer to it simply as σ.

The i-th component of the model of (1.54) is shown graphically in Fig. 1.8. Along the lines of PPR or a nonlinear form of LVR, the model consists of two parts:

1) A *linear* projection $t_i = \mathbf{w}_i^T \mathbf{x} = \sum_{j=1}^s w_{ij} x_j$ from the s-dimensional input space to one-dimensional latent variable (LV) t_i lying along the projection vector $\mathbf{w}_i$. The projection vector is chosen to be the one that is most important for explaining the behavior in the modeled output.

2) A *nonlinear* function $g_i(t_i)$ defined over this one-dimensional LV. This nonlinear function is a combination of q sigmoids, as in (1.53).

This is essentially a 3-layer perceptron with only one input, one hidden layer with q sigmoidal nodes, and one output g_i.

Together, these two components define one ridge function. This representation of the ridge function allows us to interpret it as a neural network. Hence, we can draw upon the theory and algorithms from the domain of statistical inference using neural networks to compute the best model. Let us define a sampling version of the residue for the i-th LV, similar to (1.33).

$$e_{i,j} = y_j - \sum_{k=1}^{i} g_k(\mathbf{w}_k^T \mathbf{x}_j) = e_{i-1,j} - g_i(\mathbf{w}_i^T \mathbf{x}_j). \tag{1.59}$$

This is the value of the residue for the j-th sample point, after extracting the i-th ridge function. Then, we define our model fitting criterion for the i-th ridge function, similar to (1.34), as follows:

$$(\mathbf{w}_i, g_i) = \arg\min_{\mathbf{w}, g} \sum_{j=1}^{n} |e_{i-1,j} - g(\mathbf{w}^T \mathbf{x}_j)|^2. \tag{1.60}$$

Thus, the i-th ridge function is chosen so as to minimize the least squared error in fitting the residue at that iteration across the training set. More specifically to SiLVR, using (1.53), we can write this objective as

$$(\mathbf{w}_i, \mathbf{a}_i, \mathbf{b}_i, \mathbf{c}_i) = \arg\min_{\mathbf{w}, \mathbf{a}, \mathbf{b}, \mathbf{c}} \sum_{j=1}^{n} \left[e_{i-1,j} - \sum_{k=1}^{q} a_k \sigma(b_k \mathbf{w}^T \mathbf{x}_j + c_k) \right]^2. \tag{1.61}$$

We know from Theorem 1.6 that the best g_i along any given $\mathbf{w}_i$ is the expectation of the residue at that iteration, along $\mathbf{w}_i$. However, it is not easy to compute this expectation using arbitrary training points. Hence, we do not explicitly state this constraint in the optimization formulation and assume that a good optimizer will converge close to this optimum. We make a similarly reasonable assumption for the best $\mathbf{w}_i$ (given by Theorem 1.7).

The basic training algorithm for SiLVR is shown as Algorithm 1.4. As expected, there is close similarity in the basic steps of the algorithm with the original PPR algorithm (Algorithm 1.3). The primary differences are in the representation of the ridge function, the formulation of the objective function of the "best" ridge function and the search algorithm for the "best" ridge function. The straightforward least-squares formulation of (1.61) can lead to overfitting issues of the type discussed in Sect. 1.5.1, and we absolutely must avoid such problems to achieve a well-behaved

Algorithm 1.4 The top-level SiLVR training algorithm

Require: some fixed q, the number of sigmoids per LV
1: normalize the training points $\mathbf{x}_j, y_j{}_{j=1}^n$ to mean 0 and variance 1
2: $e_j \leftarrow y_j, j = 1, \ldots, n$
3: **for** $i = 1$ to r **do**
4:     find the "best" ridge function of the form (1.53) to approximate e_j across all training points. This involves solving the objective function (1.61) appended with some penalty function to reduce overfitting (e.g., see (1.81))
5:     $e_j \leftarrow e_j - \sum_{j=1}^q a_{ij}\sigma(b_{ij}t + c_{ij}), j = 1, \ldots, n$
6: **end for**
7: the r-LV model is (1.54)

---

interpretable model from SiLVR. Hence, step 5 in Algorithm 1.4 introduced a modified objective to reduce overfitting. We discuss these issues in more detail in Sect. 1.6.4.

1.6.1.1 Model Complexity

Note that the number of model parameters to solve for is $s + 3q$ per LV, where q is the number of nodes in the sigmoid layer. Hence, if r is the total number of ridge functions that we use in the complete model, the number of model parameters is

$$n_p = r(s + 3q) = O(s), \qquad (1.62)$$

which increases linearly with the dimensionality, assuming that the number sigmoids per ridge function is independent of the dimensionality. This is a reasonable assumption, since q is determined by the anticipated nonlinearity of the functions along a single direction. This one-dimensional curve fitting depends on the modeled behavior and not on the dimensionality s. Also, q can be kept very small – 12 for our experiments across various circuit examples. The number of LVs, r, is also usually small in most cases – within 2 for our experiments. Hence, SiLVR can lead to very compact and yet, extremely flexible RSMs.

In the next section we review some theoretical results supporting this model formulation. For further algorithmic details on training the model and interpreting it in the context of RSM for circuits, the reader may skip forward to Sect. 1.6.4.

1.6.2 On the Convergence of SiLVR

In Sect. 1.5.2 we saw some results establishing the convergence of the PPR approach under some conditions that retain much practical relevance. These results do apply to SiLVR, but there is one extra consideration – the use of a *finite* number of sigmoids for estimating the ridge function g_i – that is not covered by them. We review some results in this section that help "plug this last hole".

Cybenko showed in [Cyb89] that any function continuous over the unit cube in s dimensions can be approximated arbitrarily well by a 3-layer perceptron (3LP). This is stated concretely in the following theorem.

THEOREM 1.10 (Cybenko [Cyb89]). *Let σ be any continuous sigmoidal function. Then finite sums of the form $\sum_{j=1}^{q} a_j \sigma(\mathbf{b}_j^T \mathbf{x} + c_j)$ are dense in $C[0,1]^s$. In other words, given any $f \in C[0,1]^s$ and $\epsilon > 0$, there is a sum of this form, for which*

$$\left| \sum_{j=1}^{q} a_j \sigma(\mathbf{b}_j^T \mathbf{x} + c_j) - f(\mathbf{x}) \right| < \epsilon, \quad \forall \mathbf{x} \in [0,1]^s. \tag{1.63}$$

In the case of SiLVR, we only need the one-dimensional case of this theorem. It says that there exists a $g_i(t)$ of the form (1.53) that can approximate any continuous univariate function over any bounded region of the real line. Hornik et al. [HSW89] showed similar density results for unbounded regions, with a finite probability distribution for the input space. This result gives us the confidence to use linear combinations of sigmoids as the ridge function approximators for SiLVR. Also, combined with Theorem 1.9 by Hall [Hal89] they suggest convergence for a practical implementation of SiLVR. In fact, Chui and Li in [CL92] have proved density results that can be directly applied to the complete model of SiLVR as in (1.54). We restate this here, relating it explicitly to SiLVR:

THEOREM 1.11 (Chui and Li [CL92]). *Assume that the set of possible projection vectors $\mathbf{w} \in \mathcal{W}$ satisfies the condition of Theorem 1.4. Then, for any function f continuous over any compact set $D \in \mathbb{R}^s$, and any $\epsilon > 0$, there exists a SiLVR model as in (1.54) such that*

$$|f(\mathbf{x}) - f_{\text{SiLVR}}(\mathbf{x})| < \epsilon, \quad \forall \mathbf{x} \in D. \tag{1.64}$$

This can be extended to handle the case of probability distributions of $\mathbf{x}$ over all of $\mathbb{R}^s$ using the arguments in Hornik et al. [HSW89].

Barron [Bar93] established bounds on the error of approximation using linear combinations of any fixed sigmoidal function. Here we only state the one-dimensional version. For a large class of functions f over some bounded set $B \in \mathbb{R}$, whose Fourier transform satisfies a finite-moment criterion (refer [Bar93]), the following holds.

$$\int_B (f(t)-g(t))^2 dP = O(1/q), \tag{1.65}$$

where $g(t)$ is a q-sigmoid approximation, as in (1.53). The $1/q$ behavior extends to s dimensions; that is, it is independent of the dimensionality. This says that for any given projection vector $\mathbf{w}$, the error in nonlinear function part of one component of the SiLVR model (Fig. 1.8) converges as $1/\sqrt{q}$, as long as the Fourier transform is bounded in the sense of [Bar93]. The finite-moment criterion essentially restricts the spread of the Fourier transform of f. This translates to restricting the "sharpness" and discontinuity in the function f. A counter-example is the Dirac delta function, which has a uniform Fourier transform and is understandably very difficult to model with any accuracy using smooth sigmoids. It is interesting to note the dimensionality-independent $1/\sqrt{q}$ convergence, similar to the dimensionality-independent convergence of standard Monte Carlo integration, as shown in Sect. 1.2.2.

According to this result, the more the number of sigmoids the better. However, in a sampling context, where we have only partial information because of a finite number of sampling points, this high model flexibility (complexity) can lead to overfitting problems. This overfitting problem is significantly exacerbated in the context of a PPR model like SiLVR, as discussed in Sect. 1.5.1. One way to counter overfitting is to reduce the model complexity by reducing the number of sigmoids q in the univariate approximation, so that the model is incapable of fitting the training sample exactly. Hence, there is this trade-off between high accuracy and less overfitting; i.e., between variance and bias in the model. A more detailed discussion of this issue in the context of PPR can be found in Sect. 1.5.1.

The interested reader can refer to several other results in the literature studying the density, convergence and construction of neural networks under different conditions [IM88][Fun89][Mha96][CLM96][Lig92][Bar89]. For now, we proceed on to discuss how we can interpret the SiLVR model in the context of response surface modeling for circuit.

1.6.3 Interpreting the SiLVR Model

The concept of latent variables behind SiLVR allows interpretations of the RSM that lead to useful insights in the context of circuits. We will look at two quantitative measures of these "designer's insight" that we can immediately extract from a 1-LV SiLVR model (i.e., $r=1$).

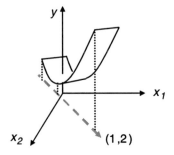

Figure 1.9. A function of two variables with dominant latent variable along $\{1,2\}$

1.6.3.1 Relative Global Sensitivity

Consider the function $y = f(x_1, x_2)$ of two variables shown in Fig. 1.9. The primary variation of f is along the shown direction $\{1,2\}$. Assuming that SiLVR can extract this feature well, the first projection vector will be given by $\mathbf{w}_1 = \{1,2\}$. The corresponding SiLVR model will be

$$\hat{y} = \sum_{j=1}^{q} a_{1j} \sigma(b_{1j}(1 \cdot x_1 + 2 \cdot x_2) + c_{1j}). \quad (1.66)$$

Hence, we can interpret from this that changes in x_2 have twice the impact on $\hat{y}$ as similar changes in x_1. In other words, $\hat{y}$ is twice as "sensitive" in a *global* sense to x_2 than to x_1. Then, if we normalize the projection vector to be of unit length ($\mathbf{w}_1/\|\mathbf{w}_1\|_2$), we can interpret the normalized projection *weights* ($w_{1j}/\|\mathbf{w}_1\|_2$) as estimates of *relative* global sensitivities of the output y to the inputs x_j. We can then define the relative global sensitivity to the j-th input variable as

$$S_j = w_{1j}/\|\mathbf{w}_1\|_2. \quad (1.67)$$

This measure of global sensitivity captures the designer's insight regarding which are the "important" variables or components in the circuit that have the most impact on the relevant circuit performance metric.

Note that these measures of global sensitivity are different from the standard measure of sensitivity $\partial f(\mathbf{x})/\partial x_j$ that models the *linear* relationship between y and the inputs x_j in a small neighborhood around a given point $\mathbf{x}$. S_j, however, takes a global view, not specific to any neighborhood around any point, but over the entire sampled input domain: it captures the overall contribution of the variable x_j to the variation in the output y. For a similar, more general interpretation of global sensitivity based on analysis of variance (ANOVA), please refer to [SK05].

Of course, S_j can be believed to be a good estimate of the global sensitivity only if a 1-LV SiLVR model explains the behavior of the circuit performance sufficiently well. In general, the accuracy of these sensitivity estimates decreases with increasing error in the 1-LV model. However, as we shall see in Sect. 1.7, a 1-LV model can extract much of the circuit behavior for some commonly used circuits. A more general definition of sensitivity using a multi-LV SiLVR model may be possible using analysis of variance, on the lines of [SK05]. This, however, is not addressed in this thesis, and can be a potential component of future work.

1.6.3.2 Input-Referred Correlation

Suppose we have 1-LV SiLVR models for two different circuit performance metrics. We can then use the global sensitivity estimates of the previous section to define a measure of correlation between the two outputs that is robust to the presence of strong nonlinearities in the relationship between them. Let us first qualitatively define the idea of "nonlinear correlation". Two variables y_1 and y_2 are strongly correlated in a nonlinear sense if they have similar *causal dependencies*. This means that the perturbations that cause changes in y_1 also cause changes in y_2.

In a circuit design context, let us consider a standard two-stage opamp. If changing the widths of the input pair of transistors causes significant changes in both the DC gain and the DC offset of the opamp, we say that the DC gain and DC offset share this causal dependence on the width of the input devices. Extending this idea, if any design change made to impact the DC gain also impacts the DC offset and vice versa, we say that the two metrics have similar causal dependencies. Note that here we are not placing any conditions on the actual relationship between the two variables. For example, Pearson's linear correlation extracts the strength of the *linear* relationship and Spearman's rank correlation (Sect. 1.6.4.1) extracts the strength of any *monotonic* relationship. Here we relax such constraints and allow any, possibly *nonlinear*, relationship. A side-effect of not assuming monotonicity is that the sign of the relationship loses meaning. For correlation measures relying on monotonicity, positive correlation means that y_1 and y_2 increase together and negative correlation means that one decreases when the other increases. However if we do not have monotonicity then both behaviors might be seen for the same pair of variables (e.g., y_1 is linear while y_2 is quadratic, but both have similar causal dependencies).

1-LV SiLVR models for y_1 and y_2 allow us to extract this measure of "nonlinear correlation" using *input-referred correlation* or IRC, defined as follows.

$$R(y_1, y_2) = \mathbf{S}^{(1)} \cdot \mathbf{S}^{(2)} = \frac{\mathbf{w}_1^{(1)} \cdot \mathbf{w}_1^{(2)}}{\|\mathbf{w}_1^{(1)}\|\|\mathbf{w}_1^{(2)}\|}, \qquad (1.68)$$

where $\mathbf{S}^{(i)}$ is the vector of s relative global sensitivities (1.67) for y_i, and $\mathbf{w}_1^{(i)}$ is the corresponding first projection vector. Thus, IRC between y_1 and y_2 is the dot product of the relative global sensitivity vectors, or the normalized first projection vectors, of y_1 and y_2. Qualitatively, the IRC value is high if y_1 and y_2 are similarly sensitive to the same set of input variables. IRC can be useful for circuit design since it quantitatively captures the designer's insight regarding the dependencies between different performances in the circuit. Such insight can help guide designers to make well-informed design decisions that do not ignore significant trade-offs.

1.6.4 Training SiLVR

The last, but arguably the most important, piece of the SiLVR RSM methodology that we have not discussed yet is the training algorithm: how do we compute all the model parameters efficiently to achieve a near-optimal model, given a finite set of training points? We hinted at the relevant issues in Sect. 1.6.1. We now discuss these in detail. The search algorithm used to find the "best" ridge function in the SiLVR training algorithm (Algorithm 1.4) has to satisfy the following three important requirements.

1) **Good generalizability**: This means that the search should strive to minimize overfitting the training points and the influence from directions orthogonal to the candidate projection vector, as discussed in detail in Sect. 1.5.1. This is accomplished by using a variable bandwidth smoothing kernel in the original PPR algorithm [FS81], as in general nonparametric methods [Pra83]. However, these techniques are not directly applicable to parametric methods like neural networks.

2) **Robust convergence**: The search should consistently settle on the same, or almost the same, model every time it is run. The desired property here is that we should be able to run the training algorithm just once and rely on the result, knowing that it is very unlikely that the search will settle in some deeply inferior local minimum.

3) **Fast convergence**: While ensuring the previous two requirements, the search should not sacrifice too much in terms of speed and the training time should be reasonable (e.g., up to several seconds).

The following four techniques are used during step 5 of Algorithm 1.4 to ensure one or more of these requirements:

1) Initialization of projection vectors using **Spearman's rank correlation** [PFTV92]. This helps start the search closer to the optimal projection vector than just a random initialization, helping achieve robust and fast convergence (requirements 2 and 3).

2) The **Levenberg–Marquardt** algorithm [Mar63] is used as the search algorithm. This algorithm blends the fast Gauss–Newton method with the robust steepest descent method to achieve fast convergence (requirement 3).

3) **Bayesian regularization** [Mac92] is used to reduce model complexity by restricting the values of the model parameters. This helps reduce overfitting and meet requirement 1.

4) A modified 5-fold **cross-validation** method is used to achieve a robust model that does not overfit the training data and, thus, helps satisfy requirements 1 and 2.

We will now discuss each of these techniques in some detail.

1.6.4.1 Initialization Using Spearman's Rank Correlation

We saw in Sect. 1.6.3.1 that the normalized projection weights of the first LV can be interpreted as relative global sensitivities of the output to the inputs. We can extend this same interpretation to the i-th LV. The normalized i-th projection weights $\{w_{ij}\}_{j=1}^{s}$ can be interpreted as relative global sensitivities of the residue being modeled by the i-th ridge function. Hence, if we can initialize the projection weights with some simple estimates of the relative global sensitivities, we can start the search closer to the global optimum, at least in the sub-space of projection weights (the entire search space has all the model parameters $(\mathbf{a}_i, \mathbf{b}_i, \mathbf{c}_i, \mathbf{w}_i)$ as dimensions). The simple estimates we use here are the Spearman's rank correlation coefficients between the output and the different inputs. Spearman's rank correlation [PFTV92] between two variables x, y, given the sample set $\{x_j, y_j\}_{j=1}^{n}$, is given by

$$\rho_S(x,y) = \frac{\sum_{j=1}^{n}(P_j - \bar{P})(Q_j - \bar{Q})}{\sqrt{\sum_{j=1}^{n}(P_j - \bar{P})^2}\sqrt{\sum_{j=1}^{n}(Q_j - \bar{Q})^2}}, \qquad (1.69)$$

where P_j and Q_j are the *ranks* of x_j and y_j in the sample set, as shown by the example in Table 1.1. To compute the rank of, say x_j, we sort all the x values in increasing order and take the position of x_j in this

x	P	y	Q
0.1	2	101	2
-0.1	1	89	1
0.89	4	130	3
0.76	3	132	4

Table 1.1. Example illustrating the concept of ranks for Spearman's rank correlation. The rank of a value is its position in a sorted list of its class; for example, 0.76 is third in the list of x values sorted in increasing order

sorted list as its rank. $\bar{P}$ and $\bar{Q}$ denote the means of the ranks. Hence, ρ_S is just Pearson's linear correlation on the ranks. However, this measure of correlation does not assume linearity like the latter, and, hence, gives better estimates of the sensitivities. It does assume a monotonic relationship between x and y. $\mathbf{w}_i$ is then initialized as the normalized vector of rank correlations between the inputs and the current residue.

$$\mathbf{w}_i = \frac{\{\rho_S\}}{\|\{\rho_S\}\|}, \quad \{\rho_S\} = \{\rho_S(x_1, e), \ldots, \rho_S(x_s, e)\}. \tag{1.70}$$

If the actual relationship is non-monotonic, in the worst case the rank correlation will not capture it and the initialization will be similar to starting at the origin.

1.6.4.2 The Levenberg–Marquardt Algorithm

We use the Levenberg–Marquardt [Mar63] algorithm to search for the best ridge function in step 5 of Algorithm 1.4. We refer to it as simply LM. LM has been found to be especially well suited for training neural networks with a least squared error formulation, as in [HM94]. It employs a blend of the fast, but sensitive Gauss–Newton method and the robust, but slower steepest descent. Steepest descent takes steps along the direction of maximum slope of the objective function. Gauss–Newton uses a quadratic approximation of the local region around the current point in the search space to estimate the minimum point. The same procedure is repeated from this new point in the next iteration. Gauss–Newton is a simplified version of Newton's method, and, like Newton's method, shows very desirable quadratic convergence close to the minimum point. However, for nonconvex surfaces, Gauss–Newton can get lost far from the global minimum. In such a situation steepest descent is a better choice.

We now delve briefly into the mathematical details of LM. Let us denote any point in our n_p-dimensional model parameter search space as **p**. Also denote the objective function to be minimized by f for this discussion on LM. The steepest descent method is an iterative procedure that traces a sequence of points $\mathbf{p}_i$, ideally towards the desired minimum point $\mathbf{p}_{min}$. The move from a current point $\mathbf{p}_{i-1}$ to the next point $\mathbf{p}_i$ is called a *step*. Steepest descent takes steps of the form

$$\mathbf{p}_i = \mathbf{p}_{i-1} - \delta \nabla f(\mathbf{p}_{i-1}), \qquad (1.71)$$

where $\nabla f(\mathbf{p})$ is the gradient vector of f at the point $\mathbf{p}$, composed of the partial derivatives of f with respect to the model parameters,

$$\nabla f = \left\{ \frac{\partial f}{\partial p_1}, \ldots, \frac{\partial f}{\partial p_{n_p}} \right\}^T, \qquad (1.72)$$

and δ is a step-size parameter. Hence, with each step, the search moves in the direction of decreasing f. As indicated by the step equation, the step length becomes smaller with smaller gradients, as happens close to a minimum point ($\nabla f(\mathbf{p}) = \mathbf{0}$). As a result, steepest descent performs well far from the optimum, but is a bad choice when the search is close to the optimum. The asymptotic convergence of steepest descent is linear.

Standard Newton's method speeds up the convergence by also using second-order information. The Newton step is given by

$$\mathbf{p}_i = \mathbf{p}_{i-1} - [\nabla^2 f(\mathbf{p}_{i-1})]^{-1} \nabla f(\mathbf{p}_{i-1}), \qquad (1.73)$$

where $\nabla^2 f(\mathbf{p}_{i-1})$ is the Hessian matrix composed of second-order partial derivatives of f with respect to the model parameters p_j. $\mathbf{p}_i$ here is basically the minimum point of a quadratic model of f around the previous iteration point $\mathbf{p}_{i-1}$.

Now, suppose that the function f is a least-squares objective function,

$$f(\mathbf{p}) = \sum_{j=1}^{n} \epsilon_j^2(\mathbf{p}), \qquad (1.74)$$

where $\epsilon_j(\mathbf{p})$ is the error for the j-th sample. An example is the SiLVR objective function in (1.61). By simple differentiation we get

$$\nabla f(\mathbf{p}) = J^T(\mathbf{p}) \mathbf{e}(\mathbf{p}), \qquad (1.75)$$

$$\nabla^2 f(\mathbf{p}) = J^T(\mathbf{p}) J(\mathbf{p}) + \sum_{j=1}^{n} \epsilon_j(\mathbf{p}) \nabla^2 \epsilon_j(\mathbf{p}), \qquad (1.76)$$

where J is the Jacobian matrix function

$$J = \begin{bmatrix} \frac{\partial \epsilon_1}{\partial p_1} & \frac{\partial \epsilon_1}{\partial p_2} & \cdots & \frac{\partial \epsilon_1}{\partial p_{n_p}} \\ \frac{\partial \epsilon_2}{\partial p_1} & \frac{\partial \epsilon_2}{\partial p_2} & \cdots & \frac{\partial \epsilon_2}{\partial p_{n_p}} \\ \vdots & \vdots & \ddots & \vdots \\ \frac{\partial \epsilon_n}{\partial p_1} & \frac{\partial \epsilon_n}{\partial p_2} & \cdots & \frac{\partial \epsilon_n}{\partial p_{n_p}} \end{bmatrix}. \quad (1.77)$$

The last term in (1.76) is very small near the solution, since $\epsilon_j(\mathbf{p})$ is very small there by definition, and it can be assumed to be ≈ 0. Then, we can write the Newton step of (1.73) as

$$\mathbf{p}_i = \mathbf{p}_{i-1} - [J^T(\mathbf{p})J(\mathbf{p})]^{-1} J^T(\mathbf{p})\mathbf{e}(\mathbf{p}). \quad (1.78)$$

This is the *Gauss–Newton* method. Note that this modification of Newton's method requires no explicit computation of second order derivatives. Newton's method has quadratic convergence near the solution because it uses the second-order information in the Hessian, and the Gauss–Newton method also shows this convergence behavior. This is significantly faster than the steepest descent method. Hence, the Gauss–Newton method is preferable once the search is close to the solution. However, far from the solution, the Hessian can be ill-conditioned, or the quadratic model might be a bad approximation of the surface, causing the search to get "lost" and move away from the actual minimum. Here, the steepest descent method, which is insensitive to the second-order behavior of the surface, is more robust and is preferable.

Recognizing this, the LM step is a intuitive blend of the two

$$\mathbf{p}_i = \mathbf{p}_{i-1} - [J^T(\mathbf{p})J(\mathbf{p}) + \mu \mathbf{I}]^{-1} J^T(\mathbf{p})\mathbf{e}(\mathbf{p}), \quad (1.79)$$

where μ is an adaptive parameter: larger μ causes steepest descent steps, while smaller μ causes Gauss–Newton steps. Note that larger μ effectively improves the conditioning of the Hessian approximation $[J^T(\mathbf{p})J(\mathbf{p}) + \mu \mathbf{I}]$ by imposing diagonal dominance [GL96]. μ is multiplied by some factor $\beta \gg 1$ (increased) when a step results in an increase in f, and divided by β (decreased) when the step reduces f. In the former case, the steepest descent part of (1.79) is increased and in the latter, the Gauss–Newton part is increased. For further details, please refer to [HM94]. We can see the flavor of typical model-trust region methods [DS96] where good solutions from the quadratic model lead to increasing the belief in the quadratic model and bad solutions lead to decreasing the belief. In fact, the LM method can be developed as a model-trust region method [DS96]. Significant improvement over steepest descent or Gauss–Newton has been seen while using LM for neural network training, as shown in [HM94].

Apart from these desirable features of LM, we also note that computing the partial derivatives for J in (1.77) is very simple in the specific case of SiLVR, because of the easy derivate calculation for the tanh sigmoid, shown in (1.58). Using (1.61),

$$\frac{\partial \epsilon_j}{\partial p_i} = \frac{\partial}{\partial p_i}\left[e_j - \sum_{k=1}^{q} a_k \sigma(b_k \mathbf{w}^T \mathbf{x}_j + c_k)\right] = \sum_{k=1}^{q} \frac{\partial}{\partial p_i}[a_k \sigma(b_k \mathbf{w}^T \mathbf{x}_j + c_k)], \quad (1.80)$$

where p_i is one of a_k, b_k, c_k for some $k \in \{1, \ldots, q\}$. Note that we have dropped the subscript for the LV here. Hence, the only derivative we need to compute is for σ, which is easily done using (1.58).

1.6.4.3 Bayesian Regularization

Optimizing the objective function in (1.61) will drive the search towards a ridge function that exactly fits the sample points along the projection vector. As discussed in Sect. 1.5.1, this is not desirable for achieving a generalizable PPR model with low overfitting. *Regularization* is a standard technique used to constrain the model complexity and reduce this overfitting behavior and involves adding a penalty term to the standard least squared error objective. Roughly speaking, the penalty term models the model complexity, using the fitting parameters themselves. If we denote the pure data-driven standard objective of (1.61) by E_D, regularization augments it as follows.

$$\min_{\mathbf{p}} E_R, \quad E_R = \beta E_D + \alpha E_p, \quad (1.81)$$

where E_p is the sum of squares of the network parameters,

$$E_p = \mathbf{p}^T \mathbf{p} = \|\mathbf{w}_i\|_2^2 + \|\mathbf{a}_i\|_2^2 + \|\mathbf{b}_i\|_2^2 + \|\mathbf{c}_i\|_2^2, \quad (1.82)$$

and α and β determine the trade-off between accuracy and generalizability, or variance and bias, respectively. Restricting the values of the network parameters reduces the flexibility in the model and increases the smoothness of the response. This is analogous to increasing the bandwidth in kernel smoothing methods [Pra83], as in the original PPR algorithm of [FS81]. Such a penalty is also known as a *roughness penalty*, and has been studied by several authors, for example [Mac92][GJP95][Bar89][HTF01].

A typical problem is estimating the proper values for α and β. A Bayesian formulation of this problem allows elegant, adaptive computation of these weights, as shown by MacKay in [Mac92]. The argument for this formulation is as follows. We recognize that the optimal values for α and β are determined by the specific neural network (or any other model)

Algorithm 1.5 Bayesian regularization in the Levenberg–Marquardt framework

1: initialize the network parameters normally, and set $\alpha = 0$ and $\beta = 1$
2: take one LM step (1.79) to minimize E_R
3: use the Gauss–Newton approximation for the Hessian, from LM

$$\nabla^2 E_R \approx 2\beta J^T J + 2\alpha \mathbf{I}_{n_p}, \quad (1.85)$$

where J is as in (1.77) and $\mathbf{I}_{n_p}$ is the $n_p \times n_p$ identity matrix, to compute $\gamma = n_p - 2\alpha/tr(\nabla^2 E_R)$

4: compute new estimates of α, β using the current point $\mathbf{p}_i$, and E_R in (1.84)
5: **if** converged as per LM criterion **then**
6:     return
7: **else**
8:     go to step 2
9: **end if**

---

structure M, and the available training data D. Given D and M, the posterior probability of some α, β is given by Bayes' rule as

$$P(\alpha, \beta | M, D) = \frac{P(D|\alpha, \beta, M) P(\alpha, \beta | M)}{P(D|M)}. \quad (1.83)$$

Under a Bayesian framework we want to use those values for α, β that maximize this probability. If we assume a prior density $P(\alpha, \beta | M)$, this can be achieved by maximizing $P(D|\alpha, \beta, M)$. $P(D|\alpha, \beta, M)$ is the likelihood of seeing the training data D, given M and some α, β. Let $\mathbf{p}^*$ denote the best choice of parameters, that minimizes (1.81). Under assumptions of Gaussian prior distributions for noise in the training set, and for the network parameters, it can be shown [FH97] that the optimum values of α, β at $\mathbf{p}^*$ are

$$\alpha_0 = \frac{\gamma}{2E_p(\mathbf{p}^*)}, \quad \beta_0 = \frac{n-\gamma}{2E_D(\mathbf{p}^*)}, \quad \gamma = n_p - \frac{2\alpha_0}{\text{trace}(\nabla^2 E_R^*)}, \quad (1.84)$$

where $\nabla^2 E_R^*$ is the Hessian of the regularization objective function (1.81) at $\mathbf{p}^*$. γ is called the effective number of parameters and is a measure of the number of model parameters actually used for reducing the error. [FH97] showed how this elegant formulation fits with the same elegance in the LM framework. The resulting algorithm for LM is shown as Algorithm 1.5, where the Hessian approximation from LM is used for $\nabla^2 E_R$.

Algorithm 1.6 Modified 5-fold cross-validation used to reduce overfitting and avoid local optima

Require: a training set D of n points
1: divide D into 5 random, nonoverlapping sets $\{D_1, \ldots, D_5\}$ which $n/5$ points each – $D = \bigcup D_i$
2: $E^* = \infty$
3: **for** $i = 1$ to 5 **do**
4: $\quad M \leftarrow$ 1-LV SiLVR model trained on $D \setminus D_i$
5: $\quad E \leftarrow$ sum of squared error of M_i on D_i
6: $\quad$ **if** $E < E^*$ **then**
7: $\quad\quad E^* = E$, $M^* = M$
8: $\quad$ **end if**
9: **end for**
10: **return** M^*

1.6.4.4 Modified 5-Fold Cross-validation

Even with this regularization technique, we cannot be completely confident of the accuracy of the resulting model on unseen test data, since we are only optimizing for the training data. Also, the surface of the objective function in (1.81) can be nonconvex because it is defined by a sum of sigmoids. Hence, there are chances that the search may settle on a local optimum that is much worse than the desired global optimum (or, at least, a very good local optimum). A popular technique used for selecting a model that is generalizable is *k-fold cross-validation*. For example, cross-validation may be used for selecting the number of sigmoids to be used in a neural network to achieve the best generalizability. For details on how it is used for model selection, the reader may refer to [HTF01]. In our case, however, the model structure is fixed. We can still exploit cross-validation to address the two issues mentioned above by choosing good model parameter values.

The modified 5-fold cross-validation that we use is shown as Algorithm 1.6. The algorithm trains 5 different SiLVR models, each time excluding one of the subsets D_i, then computes the testing error for each model on its unseen D_i, and finally, picks the model that has the lowest testing error. Hence, it ensures that testing error is used as the criterion for parameter selection, rather than only training error (with regularization). Furthermore, it runs 5 different training runs, significantly increasing the chances of finding a model close to the global optimum. Note that the cross-validation Algorithm 1.6 is run once for each LV – it is part of step 5 in Algorithm 1.4.

1.7 Experimental Results

SiLVR was implemented in Matlab. We now present some experimental results to demonstrate the performance of this implementation. We first test it on our example of (1.20), reproduced (once more) here in its general form, for the convenience of the reader.

$$y_2 = x_1 x_2 = 0.25(x_1 + x_2)^2 - 0.25(x_1 - x_2)^2$$
$$= (4ab)^{-1}[(ax_1 + bx_2)^2 - (ax_1 - bx_2)^2]. \qquad (1.86)$$

We sampled 1000 values of x_1 and x_2 from a standard normal distribution $\mathcal{N}(0,1)$, and trained SiLVR on the resulting set, using all the techniques of Sect. 1.6.4. The results are shown in Fig. 1.10. Figure 1.10 shows the training points and the surface extracted by a 2-LV SiLVR model. Comparing, with Fig. 1.4 we see that SiLVR does extract a reasonable approximation of the underlying assumption from the training set it is provided. We do see some artifacts in the under-sampled regions, but that is because of the lack of sufficient data there, and the heavy smoothing imposed on the training algorithm. Typically the under-sampled regions are less important because events occur rarely there, and errors in the model can be tolerated there. The well-sampled regions, however, must be modeled well and SiLVR does meet this criterion. Figures 1.10(b) and 1.10(c) show the first and second ridge functions extracted, respectively (along the first and second projection vectors, respectively). Again, we see agreement in shape and alignment with the solutions in Fig. 1.4. From (1.86), we know that two candidate projection vectors are $\mathbf{w}_1 = (a,b)$ and $\mathbf{w}_2 = (a,-b)$ for $ab \neq 0$. The vectors that SiLVR extracts are $\mathbf{w}_1 = (0.7285, 0.6851)$ and $\mathbf{w}_2 = (0.7296, -0.6838)$, which are very close to the expected results. This implies that even if there are errors in the model in under-sampled regions, we should still obtain good estimates of the dominant projection vectors, and hence of the relative global sensitivities and IRC, when applicable. The average absolute error on the training set is 2.23%.

Now we show results for three realistic circuit test cases, each representing a different family of circuit behavior:

1) Master–slave flip-flop with the scan chain component,

2) Two-stage RC-compensated opamp, and

3) Sub-1 V bandgap voltage reference in CMOS.

The number of process parameters range from 13 to 122 (including one inter-die parameter). SiLVR is able to extract good estimates of the LVs, along with the accompanying ridge function model, using 1,000 training

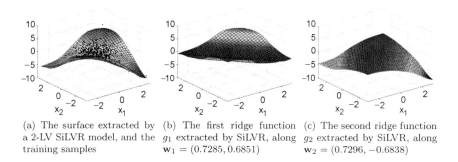

(a) The surface extracted by a 2-LV SiLVR model, and the training samples
(b) The first ridge function g_1 extracted by SiLVR, along $\mathbf{w}_1 = (0.7285, 0.6851)$
(c) The second ridge function g_2 extracted by SiLVR, along $\mathbf{w}_2 = (0.7296, -0.6838)$

Figure 1.10. A 2-LV SiLVR model for $y = x_1 x_2$ for $(x_1, x_2) \sim \mathcal{N}(\mathbf{0}, \mathbf{I}_2)$

samples for each case. The training points are generated using standard Monte Carlo sampling, following the probability distributions for the statistical parameters. The circuit simulator used is Spectre® [Kun95] by Cadence Design Systems. We also compare SiLVR with a straightforward Matlab® implementation of a near-optimal, reduced quadratic model, built using the PROBE [LLPS05] algorithm, discussed in Sect. 1.2.3. The best PROBE results (up to rank 10) are used for graphical comparisons. All models are evaluated on a separate test set of 10,000 Monte Carlo samples. Training points where the circuit does not function, are not used for modeling, but no extra samples are simulated to replace them.

1.7.1 Master–Slave Flip-Flop with Scan Chain

The first test case is a commonly seen master–slave flip-flop with scan chain shown in Fig. 1.11, which we refer to as simply MSFF. The circuit has been implemented using the 45 nm CMOS Predictive Technology Models of [ZC06]. The variations considered are random dopant fluctuation (RDF) for all transistors and one global gate oxide thickness (t_{ox}) variation. The RDF is modeled as normally distributed independent threshold voltage (V_t) variation:

$$\delta V_t \sim \mathcal{N}\left(0, \left(\frac{13.5 V_{t0}}{\sqrt{WL}}\right)^2\right), \quad (1.87)$$

where W, L are the transistor width and length in nm, and V_{t0} is the nominal threshold voltage. This results in about 30% standard deviation for a minimum-sized transistor. This is large for current CMOS technologies, but we want to make sure that SiLVR is powerful enough for future technologies too, where large variations will be inevitable. The standard deviation for t_{ox} is taken as only 2% of the nominal value, since t_{ox} is

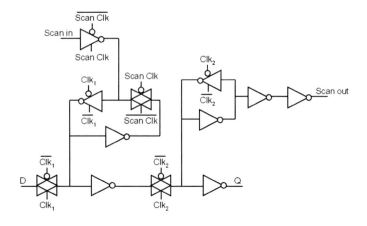

Figure 1.11. A master–slave flip-flop with the scan chain component

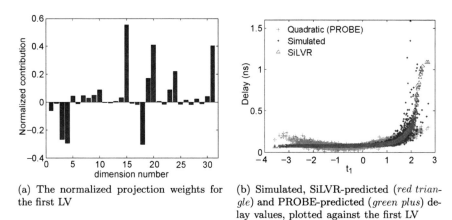

(a) The normalized projection weights for the first LV

(b) Simulated, SiLVR-predicted (*red triangle*) and PROBE-predicted (*green plus*) delay values, plotted against the first LV

Figure 1.12. Performance of SiLVR on the MSFF test case

typically better controlled than RDF. The number of statistical parameters, or input dimensionality of the model, is 31. We are modeling one output: the clock-output delay of the flip-flop (τ_{cq}). The setup time of the flip-flop is such that the variations result in the onset of some metastable behavior, resulting in strongly nonlinear behavior in some parts of the sampled region of the statistical parameter space. This realistic situation makes the modeling problem a harder test case.

Figure 1.12(a) shows the projection vector $\mathbf{w}_1$ for the first extracted latent variable t_1, and Fig. 1.12(b) plots the simulated and predicted delay values against t_1. The latter shows the predictions from SiLVR and also from the best reduced quadratic model. We can clearly observe two things: 1) only 6–8 out of 31 input dimensions (corresponding to

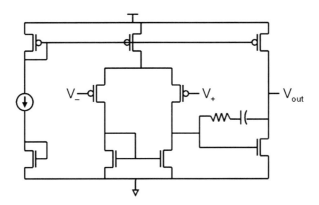

Figure 1.13. A 2-stage RC-compensated operational amplifier

transistors in the circuit) affect the output, and 2) SiLVR performs much better than a quadratic model. Also, note that just using one LV we are able to explain the general behavior of this test case. Figure 1.21 (a few pages later) shows error bars comparing the error of SiLVR against the error of PROBE, and we can immediately see the improvement in modeling accuracy. These are errors for the best SiLVR model and the best PROBE model: both PROBE and SiLVR perform best for rank-one and a one-LV model, respectively. Beyond this, we see only overfitting of the training data and larger errors on the test data. Table 1.3 shows this in more detail, as it compares the errors quantitatively with increasing number of LVs/rank: the best average error is reduced by 2.5×: from 16.3% for PROBE to 6.4%. The results for rank (number of LVs) greater than 6 do not provide any relevant insight and are excluded to avoid clutter.

1.7.2 Two-Stage RC-Compensated Opamp

This next test case [GJLM01], shown in Fig. 1.13, is representative of a large class of circuits in the *analog* domain: amplifiers. We test SiLVR on the DC, AC and transient characteristics of the opamp. The opamp has been implemented using models from the Cadence 90 nm Generic PDK library. Once again, we model RDF on all transistors as independent variation on the threshold voltage V_t. We also include a global t_{ox} variation, and variations on the passives (resistors, capacitors) and the current source. All variations are assumed to be normally distributed. The V_t standard deviation is about 18% of nominal V_t:

$$\delta V_t \sim \mathcal{N}\left(0, \left(\frac{5 \text{ mV}}{\sqrt{WL}}\right)^2\right), \quad (1.88)$$

where W, L are the transistor width and length in μm. The standard deviation for t_{ox} is taken as 2% of the nominal value. Each passive and current source component has its own normally distributed variation with a standard deviation of 5%. The resulting input dimensionality for the RSM is 13, and we are modeling five performance metrics (outputs):

1) DC gain

2) Unity gain frequency (UGF)

3) Phase margin (PM)

4) Settling time (ST)

5) DC input offset

For details regarding these metrics and opamp operation, please refer to any standard textbook on circuit design, e.g., [GJLM01].

Figure 1.14 shows the first projection vector $\mathbf{w}_1$ for each of the outputs. Figure 1.15 plots the simulated and predicted values for each output against the respective first LV t_1, on the test points. The following observations are obtained immediately from these figures:

- Strongly nonlinear behavior exists even for simple circuits like the 2-stage opamp: for DC gain and PM in this case.

- A quadratic model performs well for near-linear behaviors, as expected, but has large errors for these strongly nonlinear behaviors.

- One (the first) LV is able to explain much of the behavior for some circuits: both the MSFF and the opamp till now.

- SiLVR is able to model even the strongly nonlinear behaviors reasonable well.

Apart from these obvious ones, we make some more subtle, but important observations. Having the explicit projections, as a result of the projection pursuit approach, provides deep insight into the circuit behavior. First, we can actually see the behavior clearly, removing any need for guesswork. We found a direct application of this advantage during the course of performing these experiments. Our initial results for the input offset showed a surprising (and, as it turned out, erroneous) step-shaped behavior when the offset values were plotted against the first LV extracted by SiLVR. This is shown in Fig. 1.16. This result can also be found in our initial publication [SR07a]. However, this step behavior was unexpected for the offset of the opamp: we expected a near-linear behavior as described in [GJLM01]. This led to further investigation, resulting

SiLVR

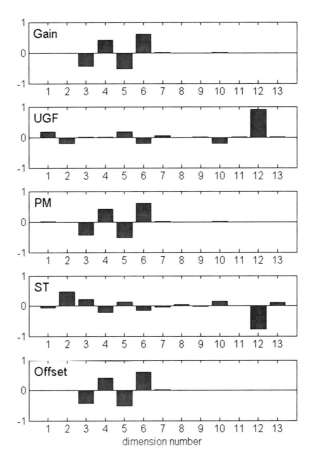

Figure 1.14. Opamp test case: Normalized projection vectors for the first LVs of the opamp metrics: we can see the strong relationship between gain, PM and offset

in the discovery of the cause of this anomaly – a tolerance parameter in the circuit simulator was too loose. On correcting the parameter, and re-running SiLVR training, we obtained the expected near-linear behavior shown in Fig. 1.15. This example shows just one of may possible scenarios where the better modeling flexibility, visualization potential and interpretive power that SiLVR provides can be practically useful. The step behavior could be observed easily because we could reduce the dimensionality to the most important one and visualize the behavior easily.

SiLVR also provides us some quantitative measures to better understand circuit behavior: relative global sensitivities and input-referred correlation (Sect. 1.6.3). If we look at the projection vectors for gain, PM and offset in Fig. 1.14, we can immediately see that these outputs de-

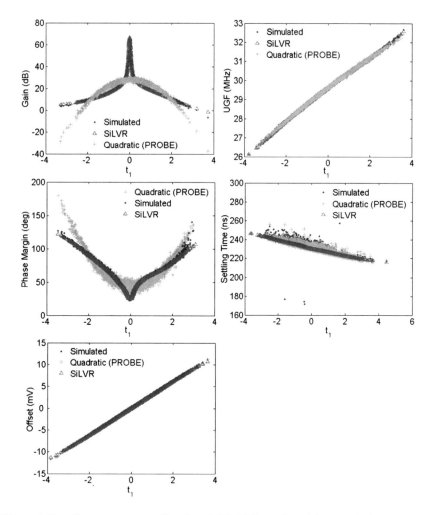

Figure 1.15. Opamp test case: Simulated, PROBE-predicted (*green plus*) and SiLVR-predicted (*red triangle*) opamp outputs, plotted against first LV. For the nonlinear cases, the simulated and SiLVR-predicted graphs coincide in many places

Output pair	|Linear corr.|	|Rank corr.|	IRC
Gain–PM	0.871	0.986	1.000
PM–offset	0.119	0.161	1.000
Gain–offset	0.054	0.099	1.000

Table 1.2. Rank and linear correlation compared with IRC as a measure of correlation between strongly correlated opamp metrics

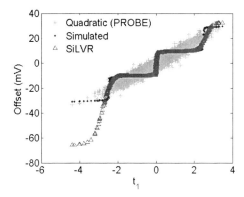

Figure 1.16. Staircase behavior for incorrectly simulated input offset modeled by SiLVR

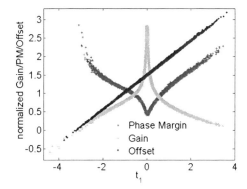

Figure 1.17. Simulated opamp gain, phase margin and offset plotted against the LV for gain, showing strong correlation among the three

pend almost identically on the same parameter subset (parameters 3–6): these are the driver and load devices in the input differential amplifier. Hence, they have similar causal dependencies, as defined in Sect. 1.6.3.2, and are strongly correlated in a nonlinear sense. This is confirmed by plotting all three simulated metrics against the first LV of *gain*, as in Fig. 1.17. This is where the power of SiLVR is really evident. Table 1.2 compares the rank correlation and linear correlation among these metrics, with the IRC. We see that the rank correlation performs better than linear correlation, but both completely fail to capture the strength of the relationship between gain and offset, and gain and PM. At the same time, IRC succeeds nicely.

Figure 1.21 compares the average absolute percentage error for SiLVR and PROBE on this test case. Quantitative comparisons of the errors

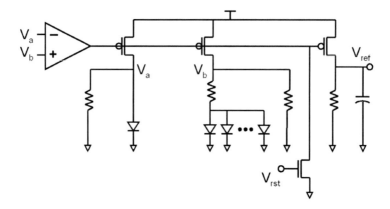

Figure 1.18. Low-voltage CMOS bandgap voltage reference circuit from [BSUM99], with a parameter space of 122 dimensions

are provided in Table 1.4, for increasing number of LVs and rank of the quadratic model. The results for rank (number of LVs) greater than 6 do not provide any relevant insight and are excluded to avoid clutter. As expected, a quadratic model has large errors for the nonlinear, non-quadratic behavior of gain and PM. SiLVR can model these well and reduce the error significantly – by up to an absolute improvement of 34% for gain. The errors for the near-linear outputs worsen a little, but are still within reasonable limits. For this test case too, a 1-LV SiLVR model has the lowest testing error amongst all SiLVR models, as we can see from the table.

1.7.3 Sub-1 V CMOS Bandgap Voltage Reference

Figure 1.18 shows a low-voltage CMOS bandgap voltage reference circuit, proposed in [BSUM99]. The circuit is able to provide reference voltages that are less than 1 V, and is built using standard CMOS technology. It was chosen for its relevance in today's and tomorrow's low-voltage designs, and also because the related RSM problem has a high input dimensionality of 122 and strong nonlinear behavior. The opamp in the circuit is the same as in Sect. 1.7.2. The circuit has 101 diodes. The transistor device and variation models are the same 90 nm CMOS as the opamp. Variations in each diode are modeled as a normally distributed variation on the saturation current, with standard deviation of 10%. Each resistor and capacitor has its own normally distributed variation source, with a standard deviation of 5%. There are a total of 121 local variation parameters and one global t_{ox} variation. In this case, we measure two metrics: 1) the output voltage V_{ref}, and 2) the dropout

SiLVR

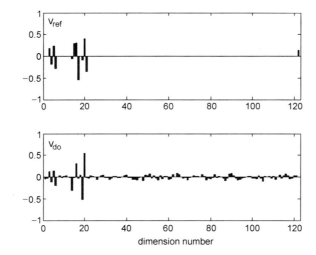

(a) Normalized projection vector for the first LVs of the voltage reference circuit metrics

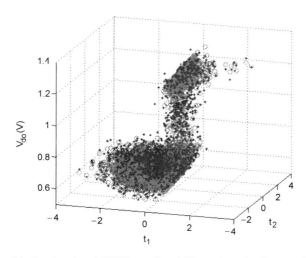

(b) Simulated and SiLVR-predicted V_{do} against the first two LVs

Figure 1.19. Performance of SiLVR on the sub-1 V CMOS voltage reference circuit test case

voltage V_{do}. V_{do} is the difference between the supply voltage and V_{ref}, when V_{ref} falls by 1% of its nominal value: lower V_{do} implies a circuit more robust to variations in the supply voltage. The nominal V_{ref} we designed for is 600 mV.

Figure 1.19(a) shows the 122-dimensional projection vector for the first LVs of the bandgap performance metrics. Figure 1.20 plots the sim-

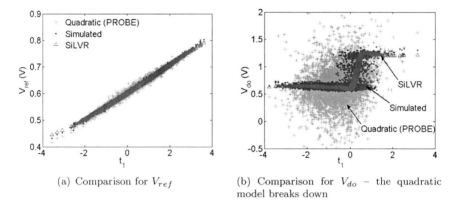

Figure 1.20. Simulated, PROBE-predicted (*green plus*) and SiLVR-predicted (*red triangle*) outputs, plotted against the first LV, for the sub-1 V CMOS voltage reference circuit test case

ulated and predicted outputs against their corresponding LVs. Here we see that PROBE performs well for the linearly behaved V_{ref}, but completely breaks down for the nonlinearly behaved V_{do}. SiLVR, however, *is* able extract a good estimate of this strong nonlinear behavior, as evidenced also by Table 1.3. For the case of V_{do}, we can actually improve the fit further by using a 2-LV model, as shown both by Fig. 1.19(b) and by the last column in Table 1.3. The former plots the simulated and SiLVR-predicted values of the dropout voltage for the test set points. Only for this figure, the SiLVR model was trained using a (separate) training set of 10,000 points to achieve a fit that is visually obvious in three dimensions. However, the results in Table 1.3 and in Fig. 1.21 are for a model trained using the standard sample size of 1,000 points.

Hence, even though we started with a large dimensionality of 122, only 2 LVs can still explain most of the behavior. Also, the normalized inner product of the first two projection vectors $\frac{\mathbf{w}_1^T \mathbf{w}_2}{\|\mathbf{w}_1\|\|\mathbf{w}_2\|}$ is only 1.2e–3, meaning that they are almost orthogonal. This implies that SiLVR can extract almost all the information from the first LV before looking at the second LV. We saw similar results supporting this inference for the example $y = x_1 x_2$ at the beginning of Sect. 1.7.

1.7.3.1 Training Time

The training run times to build each LV are quite reasonable, even with the complex cross-validation strategy to improve neural network robustness: each LV requires 13–24 CPU seconds of Matlab® computation. This is especially attractive for higher dimensional cases like the volt-

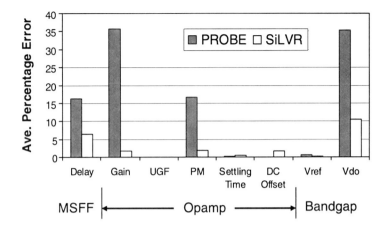

Figure 1.21. Best SiLVR errors compared with best PROBE errors: PROBE shows large errors for the nonlinear performances. SiLVR significantly reduces those errors and maintains low errors for the near-linear cases (UGF, Settling Time, DC offset, V_{ref})

r	MSFF delay		Bandgap V_{ref}		V_{do}	
	SiLVR	PROBE	SiLVR	PROBE	SiLVR	PROBE
1	6.41	16.3	0.278	0.639	11.1	35.4
2	7.24	19.7	0.341	0.707	10.4	39.9
3	8.15	21.3	0.374	0.726	11.9	41.9
4	8.17	22.2	0.375	0.736	11.9	42.3
5	8.35	22.9	0.374	0.739	11.9	42.4
6	7.79	23.2	0.374	0.738	11.9	42.4

Table 1.3. Average percentage error on a test set of 10,000 Monte Carlo samples, for MSFF and the voltage reference. r is the rank of the quadratic model or the number of LVs used in the SiLVR model, as applicable

age reference, where even the simple quadratic model of PROBE can be relatively expensive to train.

1.8 Future Work

SiLVR possesses some very desirable features as an RSM technique. It elegantly handles nonlinear surfaces, enables performance-oriented dimensionality reduction, provides useful quantitative measures to understand the circuit design problem (relative global sensitivity and IRC), and enables insightful visualization of performance behavior in much reduced dimensions. Even before its application to automatic optimization, these features can find good use in the manual design process. For

r	Opamp gain		UGF		PM		Settling time		Input offset	
	SiLVR	PROBE	SiLVR	PROBE	SiLVR	PROBE	SiLVR	PROBE	SiLVR	PROBE
1	1.69	35.7	0.061	0.042	1.90	16.7	0.507	0.248	1.62	0.58
2	1.74	36.4	0.063	0.022	1.97	16.8	0.517	0.240	1.80	0.12
3	1.73	36.9	0.066	0.013	1.94	16.9	0.528	0.243	2.11	0.10
4	1.66	37.2	0.068	0.010	1.99	16.9	0.531	0.244	2.38	0.09
5	1.69	37.1	0.070	0.009	2.01	16.9	0.540	0.244	2.41	0.09
6	1.72	37.0	0.072	0.009	2.03	16.9	0.552	0.245	2.59	0.09

Table 1.4. Average percentage error on a test set of 10,000 Monte Carlo samples, for the opamp. r is the rank of the quadratic model or the number of LVs used in the SiLVR model, as applicable

example, after running an increasingly popular Monte Carlo run for yield analysis, the designer can obtain a SiLVR model and all its by-product features in a few seconds without running any more simulations. This is a very simple use-mode that is minimally intrusive to most circuit design flows, and still provides a useful new design tool to the designer. The practical, real examples from the case of the opamp in Sect. 1.7.2 illustrate this usefulness.

Given this, much can still be done to extend the power and usefulness of SiLVR. Here we briefly introduce some possible directions of further research, targeting various aspects of this RSM strategy.

- Researchers in statistics and data mining have developed techniques to refine the PPR model. One relevant technique is so called *backfitting* [HTF01]. In this procedure, after the i-th projection in the PPR model is extracted, the previously extracted projections are re-optimized, to better model the residue after removing the behavior modeled by this i-th projection.

- The SiLVR training algorithm emphasizes the reduction of overfitting, by employing a minimal number of sigmoids and using regularization and cross-validation techniques. It is not very clear if it currently sits at the best trade-off between accuracy and generalizability; that is, between variance and bias. This issue deserves further investigation to determine a near-optimal trade-off for the SiLVR model.

- Finally, a goal of most RSM strategies is to be employed as circuit performance models in automatic yield-aware optimization. Such is also the case for SiLVR. How can SiLVR be best incorporated in yield-aware circuit synthesis? This is a "loaded" question and any inquiry into its answer will require answering several other questions: How can SiLVR be adapted to work across both the statistical parameter and design variable spaces to allow larger model-trust regions in the design space? How can the information available from SiLVR be best used for guiding the search algorithm? Which search algorithms fit best with SiLVR?

Chapter 2

Quasi-Monte Carlo for Fast Statistical Simulation of Circuits

2.1 Motivation

Continued device scaling has dramatically increased the statistical variability with which circuit designers must contend to ensure the reliability of a circuit to these variations. As discussed in the introduction to this thesis, traditional process corner analysis is no longer reliable because the variations are numerous and much more complex than can be handled by such simple techniques. Going forward, it is increasingly important that we account accurately for the statistics of these variations during circuit design. In a few special cases, we have analytical methods that can cast this inherently statistical problem into a deterministic formulation, e.g., optimal transistor sizing and threshold assignment in combinational logic under statistical yield and timing constraints, as in [MDO05]. Unfortunately, such analytical solutions remain rare. In the general case, some combination of complex statistics, high dimensionality, profound nonlinearity or non-normality, stringent accuracy, and expensive performance evaluation (e.g., SPICE simulation) thwart our analytical aspirations. This is where Monte Carlo methods [Gla04] come to our rescue as true statistical methods.

Monte Carlo simulation can emulate a real statistical process using a given technique for simulating any event from this statistical process. For example, the performance of chips coming out of a manufacturing process is emulated by simulating multiple instances of the relevant circuit, with each instance having a different set of values for its manufacturing related parameters. Over the years, Monte Carlo has become a standard technique for statistical simulation of circuits and for yield estimation during the design phase [SP81][HLT83][SKC99][Eli94]. However, we gain the flexibility and accuracy that Monte Carlo offers at the cost of speed: a single Monte Carlo run can cost a few thousand SPICE simulations. Given its importance, it is surprising that Monte Carlo has not received the research effort it deserves, from the EDA community. There has been much research for methods to either replace Monte Carlo simulation all together, using acceptance region modeling [AMH91][AGW94][SVK94], or replacing SPICE simulations, using response surface modeling [YKHT87][FD93][LLPS05]. The SiLVR model presented in Chap. 1 falls under the category of response surface models. However, these methods gain speed by sacrificing accuracy, and in many cases accuracy is very important. An ideal solution would be to somehow speed up Monte Carlo simulation, while still using SPICE simulations and maintaining the generality of its application. There has been some research on this [HLT83][SKC99], but there is much more that needs to be done.

Here we observe that similar problems exist in various other fields of science and engineering. In particular, we look at the field of computational finance, where pricing financial instruments and derivatives (e.g., Asian options, mortgage-backed securities) requires the simulation of high dimensional stochastic processes, for which Monte Carlo has remained the main practical method [Gla04]. These problems are not only very nonlinear, they can also be quite large: pricing a portfolio of options or securities over a several year horizon can create problems with 1,000+ statistical variables, as in [NT96b]. Accuracy is often required to the level of one basis point (a relative accuracy of 10^{-4}) under impressively short time constraints (minutes, in the case of real-time arbitrage). In this chapter we attempt at redepolying a particularly successful Monte Carlo technique from this domain to our problem domain of circuits. This technique is commonly referred to as *quasi-Monte Carlo* (QMC) and is essentially Monte Carlo, but using a *deterministic* set of points from some, so called, *low-discrepancy sequence*.

Note that the theoretical underpinnings of QMC are not completely new, as evidenced by number theoretic results by Halton in 1960 [Hal60a]. However, recent developments in both theory and implementation complexity, along with the empirical discovery that it is unexpectedly efficient at evaluating certain high-dimensional integrals, have propelled QMC onto the center stage in the computational finance world, as evidenced by extensive articles in both popular and practitioner literature (The Economist, August 12, 1995; The New York Times, September 25, 1995; Risk Magazine). QMC has also found application for high dimensional integration problems in physics [MC95][Spa95]. A motivational example from finance is provided by Ninomiya and Tezuka in [NT96b],

Quasi-Monte Carlo 61

where they evaluate the price of a five-year discount bond. For this 1,439-dimensional problem, they observe a speedup of about 150 for an accuracy level of 1 basis point, on using QMC instead of Monte Carlo. Much work has been done to study the application of QMC to finance problems [CMO97][ABG98][OE04][PT95]. Our goal here will be to study its application to circuit problems.

In the rest of the chapter, we will review the standard Monte Carlo method and its convergence behavior, relevant results from number theory and resulting QMC techniques, and our proposed framework for applying QMC to statistical simulation of circuits. While developing our framework we will also discuss some important idiosyncrasies of the QMC technique, because of which a naive, direct application might not work well. Based on this discussion we will then develop the essential pieces of our flow. We shall see in the results section that this proposed framework can lead to speedups of about 2 to 50 times over standard Monte Carlo while maintaining the same level of accuracy. A concise version of this chapter was presented in [SR07b].

2.2 Standard Monte Carlo

Let us first concretely define the canonical problem that Monte Carlo simulation addresses. We take two seemingly very different examples to arrive at common terminology. We will then base further discussion on this common terminology and the canonical problem.

2.2.1 The Problem: Bridging Computational Finance and Circuit Design

Consider two problems from two completely different domains:

A. pricing an Asian option in computational finance, and

B. estimating circuit yield in VLSI design.

Let us see what is common between these two problems. This will allow us to develop a canonical representation for the general problem that Monte Carlo solves, which will further enable us to clearly understand and apply related results.

2.2.1.1 Pricing an Asian Option

An *option* gives an investor the right to purchase one unit of a security at a specified *strike price* K at a future time T; for example, the right to purchase shares of company XYZ at 5 dollars per share on a fixed date in the future. Given K, T, we wish to determine the price that the investor should pay for this option at present time 0. Merton expanded the work by Black and Scholes to develop the *Black–Scholes* options pricing model [Mer73]. Merton and Scholes received the Nobel Prize in Economics in 1973 for this and other related work. Among other results, the model gives the *payoff* on an arithmetic *Asian* option, one of several types of options [Gla04], as

$$\left[\frac{1}{T}\int_0^T S(t)dt - K\right]_+, \quad \text{where } [\cdot]_+ = \max(0, \cdot), \tag{2.1}$$

where $S(t)$ is the price of the underlying security (stock) at time t. $S(t)$ is given as

$$\frac{dS(t)}{dt} = rdt + \sigma x(t)\sqrt{dt} \tag{2.2}$$

$$\Rightarrow \quad S(t) = S(0)e^{[(r-0.5\sigma^2)t + \sigma \int_0^t x(t)\sqrt{dt}]}, \tag{2.3}$$

where r is the *risk-free, continuously compounded interest rate*, and $x(t)$ is a random process such that for every instant t, $x(t) \sim \mathcal{N}(0,1)$. Thus, $W(t) = \int_0^t x(t)\sqrt{dt}$ is a Wiener process [Gla04]; that is, $W(t) \sim \mathcal{N}(0,t)$. Here, $x(t)\sqrt{dt}$ embodies the random volatility in the price of the security and σ is the magnitude of this volatility.

The Black–Scholes model gives the appropriate price of the option at time 0 as the expected value of the *discounted* payoff:

$$K_0 = E\left\{e^{-rT}\left[\frac{1}{T}\int_0^T S(t)dt - K\right]_+\right\}, \tag{2.4}$$

where e^{-rT} accounts for the fact the option will be purchased at time 0, but exercised at a future time T. The typical way to evaluate this price K_0 is to first discretize time t into s samples, with equal steps of size Δt as

$$t_0 = 0, \quad \Delta t = \frac{T}{s}, \quad t_i = t_{i-1} + \Delta t, \ i \in \{1, \ldots, s\}. \tag{2.5}$$

Then, the $x_i = x(t_i)$ are s independent, identically distributed random variables $\sim \mathcal{N}(0,1)$. We can now write the security price from (2.3) as

$$S(t_i) \approx S(0)e^{[(r-0.5\sigma^2)i\Delta t + \sigma\Delta t \sum_{j=0}^{i} x_j]}, \quad i = \{1, \ldots, s\}. \tag{2.6}$$

Then, evaluating $S(t_i)$ at each time sample, we can numerically approximate the $\frac{1}{T}\int_0^T S(t)dt$ in (2.4) as

$$\frac{1}{T}\int_0^T S(t)dt \approx \bar{S}_s = \frac{1}{s}\sum_{i=1}^{s} S(t_i) \tag{2.7}$$

and compute the option price from (2.4) as

$$K_0 \approx E\{e^{-rT}[\bar{S}_s - K]_+\}. \quad (2.8)$$

Note that $\bar{S}_s$ is a function of the s random variables $\{x_i\}_{i=1}^s$ that follow a joint multivariate density distribution $\pi(\mathbf{x}) = \mathcal{N}(\mathbf{0}, \mathbf{I}_s)$. We can then write the option price as

$$K_0 \approx \int_{\mathbb{R}^s} f(\mathbf{x})\pi(\mathbf{x})d\mathbf{x}, \quad \text{where } f(\mathbf{x}) = e^{-rT}[\bar{S}_s - K]_+. \quad (2.9)$$

Hence, the problem is now of evaluating an integral over an s-dimensional space.

2.2.1.2 Estimating Circuit Yield

Consider some circuit with s statistical parameters, or simply *inputs*, $\{x_i\}_{i=1}^s$ and s_y performance metrics, or simply *outputs*, $\{y_i\}_{i=1}^{s_y}$. The relationship between the outputs and the inputs can be written as

$$\mathbf{y} = \mathbf{f}_{sim}(\mathbf{x}) \quad (2.10)$$

where evaluating $\mathbf{f}_{sim}$ might involve running one or more circuit simulations (e.g., AC analysis) and subsequent computations to compute the metrics (e.g., gain), as needed to compute the metrics in $\mathbf{y}$. Of course, $\mathbf{f}_{sim}$ also take the design variables as arguments, but we assume a fixed design for this discussion. Also, there are some specifications that the performance metrics must meet for an acceptable design. Denoting these specifications by $\{t_i\}_{i=1}^{s_y}$, we require $\{y_i \leq t_i\}_{i=1}^s$, or equivalently $\mathbf{y} \leq \mathbf{t}$. Please note that here we use $\leq$ without any loss of generality. If for some given $\mathbf{x}$, the design meets this criterion, we denote the event as a *pass* event, otherwise it is a *fail* event. In the context of manufacturing variations, we might be interested in estimating the yield of the circuit given probability distributions for the statistical parameters. The yield is the percentage of manufactured instances of the circuit that pass the specifications. We now state this mathematically. Let us define $\mathcal{A}$, the *acceptance region* for a given design, as the set of input vectors that give us a passing circuit:

$$\mathcal{A} = \{\mathbf{x} : \mathbf{f}_{sim}(\mathbf{x}) \leq \mathbf{t}, \ \mathbf{x} \in \mathbb{R}^s\}. \quad (2.11)$$

Also, define the *characteristic function* of $\mathcal{A}$ as

$$I_{\mathcal{A}}(\mathbf{x}) = \begin{cases} 1, & \mathbf{x} \in \mathcal{A} \\ 0, & \mathbf{x} \notin \mathcal{A} \end{cases} \quad (2.12)$$

which is 1 for pass and 0 for fail. This is also known as the *indicator function* in the VLSI CAD literature [HLT83]. Now, we can define the circuit yield as the probability of a circuit instance lying in the acceptance region:

$$Y_t = P(\mathbf{x} \in \mathcal{A}) = E(I_\mathcal{A}(\mathbf{x})) \qquad (2.13)$$

which can be written as

$$Y_t = \int_{\mathbb{R}^s} I_\mathcal{A}(\mathbf{x}) \pi(\mathbf{x}) d\mathbf{x}. \qquad (2.14)$$

This is now a problem of s-dimensional integration, similar to the problem of pricing Asian options.

2.2.1.3 The Canonical Problem

Equations (2.9) and (2.14) are identical in their form

$$Q = \int_{\mathbb{R}^s} g(\mathbf{x}) \pi(\mathbf{x}) d\mathbf{x} \qquad (2.15)$$

and suggest a canonical form for the general problem. Only one step remains before we can reach this canonical form. let π_i be the marginal probability density distribution for x_i and Π_i be the corresponding marginal cumulative distribution. Then, for independent x_i, we can write (2.15) as

$$\begin{aligned} Q &= \int_{\mathbb{R}^s} g(x_1,\ldots,x_s) \pi_1(x_1) \pi_2(x_2) \ldots \pi_s(x_s) d\mathbf{x} \\ &= \int_{[0,1]^s} g(\Pi_1^{-1}(z_1),\ldots,\Pi_s^{-1}(z_s)) d\mathbf{z}, \end{aligned} \qquad (2.16)$$

leading us to the canonical form we seek by renaming z_i as x_i:

$$Q = \int_{C^s} f(\mathbf{x}) d\mathbf{x}, \quad C^s = [0,1]^s \qquad (2.17)$$

where C^s is the unit cube in s dimensions. For the rest of our discussions in this chapter we consider only C^s as our integration domain and assume that all required transformations have been incorporated into the function f.

2.2.2 Monte Carlo for Numerical Integration: Some Convergence Results

The general integration problem does not usually admit an analytical solution. A common approach to solve it then is to use numerical integration or *quadrature*, also known as *cubature* for $s \geq 2$ [Str71][Coo99].

These quadrature rules typically involve evaluating the function f at strategically placed points in C^s and doing a weighted sum to arrive at the estimate for the integral Q. The problem with these classical cubature methods is that they become intractable as the dimensionality s increases. The following theorem [Nik50] states this problem concretely.

THEOREM 2.1 ([Nik50]). *Let $f \in W_p^k(C^s)$, where $W_p^k(C^s)$ is the Sobolev class [Ada75] of functions defined on the unit cube C^s whose weak derivatives up to order k exist and are bounded under the L_p norm (see Sect. 1.4 and [Ada75] for definitions). Let $Q(f)$ be the exact integral for f, and $Q_n^{det}(f)$ be any n-point quadrature approximation to $Q(f)$. If $pk > s$ then*

$$\inf_{Q_n^{det}} \sup_{\{f : \|f\|_{k,p} \leq 1\}} |Q(f) - Q_n^{det}(f)| = \Theta(n^{-k/s}) \quad (2.18)$$

where the norm $\|f\|_{k,p}$ is the norm for the Sobolev space $W_p^k(C^s)$.

The theorem essentially says that for a given class of smooth functions, the error of any numerical quadrature method using n deterministic points decreases asymptotically as $\Theta(n^{-1/s})$ with the dimensionality. This implies that to halve the error, the number of points must increase by a factor of 2^s. Also, to maintain the same error, the number of quadrature points must increase exponentially with the dimensionality s. Thus, for circuit yield estimation, the number of circuit simulations in (2.14) must increase exponentially with the number of statistical parameters. This can very easily become intractable, even for very few parameters. Here, we have run into the well-known *curse of dimensionality*. We face this "curse" in all the three chapters of this thesis and a part of each proposed method is some technique to defeat it.

Monte Carlo is able to defeat this curse. This is the primary reason for its popular adoption for computing high-dimensional integrals in a wide variety of fields. Another class of quadrature techniques based on sparse grids proposed by Smolyak [Smo63] also improves the convergence to make moderate-dimensional integration feasible. However, for large dimensions (100s) only Monte Carlo techniques are known to be tractable. For more details on sparse grid-based quadrature, please refer to [GG98]. The quadrature points used by standard Monte Carlo are randomly chosen. We will also refer to these as sample/sampling points in the context of Monte Carlo. In general, any random method for computation is a Monte Carlo method [Hei96], but we focus primarily on independent Monte Carlo, where every point is generated independently of the other points. Examples of dependent Monte Carlo methods are Markov chain Monte Carlo methods like Gibbs sampling and simulated

Algorithm 2.1 The standard Monte Carlo algorithm

Require: function f, joint probability distribution $\Pi(\mathbf{x})$, and sample size n
1: **for** $i = 1$ to n **do**
2:     randomly generate $\mathbf{x}_i = (x_1, \ldots, x_s)$ from Π
3:     evaluate $y_i = f(\mathbf{x}_i)$
4: **end for**
5: **return** Monte Carlo estimate $Q_n = \frac{1}{n}\sum_{i=1}^{n} y_i$
---

annealing: a good survey is provided in [Fis06]. The standard Monte Carlo algorithm is shown as Algorithm 2.1.

One Monte Carlo run involves evaluating the function f at n randomly chosen locations in the input space. Since $\mathbf{x}_i$, and hence y_i, are independent and identically distributed, the Monte Carlo estimate Q_n converges almost surely to Q as the sample size n is increased by the strong Law of Large Numbers [HC71]; i.e.,

$$P(\lim_{n\to\infty} Q_n = Q) = 1. \qquad (2.19)$$

From Algorithm 2.1, we can easily see that if we ran multiple n-point Monte Carlo runs, we would obtain a different estimate Q_n each time. As a result, the integration error of Monte Carlo is probabilistic in nature and a deterministic bound, as in Theorem 2.1, does not make sense. An *average* error, however, does make sense. Bakholov [Bak59] showed the following result.

THEOREM 2.2 (Bakholov [Bak59]). *Assume the conditions of Theorem 2.1. The average Monte Carlo integration error is* $\Theta(n^{-\frac{k}{s}-\frac{1}{2}})$.

A proof can be found in [Hei94]. Thus, we can significantly improve over the exponential complexity of the worst error for deterministic methods. For small s, the convergence behavior is close that for the classical quadrature methods, $n^{-\frac{k}{s}}$. However, for moderate to large values of s (typically ≥ 6), the dimension dependent part becomes negligible and the convergence is close to $n^{-\frac{1}{2}}$. The extra gain of $n^{-\frac{k}{s}}$ is possible if we exploit the smoothness of the function using variance reduction techniques: these are enhancements to the standard algorithm that reduce the variance of the estimate Q_n [Fis06]. If we do not exploit the smoothness, or if f is not necessarily smooth, we can still derive a similar result using standard statistics.

Quasi-Monte Carlo

THEOREM 2.3. *Let $f \in L_1(C^s)$ be integrable over C^s. Define*

$$\sigma(f) = \left[\int_{C^s} (f(\mathbf{x}) - \bar{f})^2 d\mathbf{x}\right]^{\frac{1}{2}}, \quad \bar{f} = \int_{C^s} f(\mathbf{x}) d\mathbf{x} = Q. \quad (2.20)$$

Then the average (r.m.s.) error of Monte Carlo is

$$\sqrt{E[(Q - Q_n)^2]} \to \frac{\sigma(f)}{\sqrt{n}} \quad \text{as } n \to \infty. \quad (2.21)$$

PROOF. This is obvious from the central limit theorem (Theorem 3.2 in Sect. 3.2.2) [HC71], which says that

$$\lim_{n \to \infty} \frac{Q_n - Q}{\sqrt{\sigma^2/n}} \xrightarrow{d} \mathcal{N}(0, 1). \quad (2.22)$$

Hence, the Monte Carlo error decreases asymptotically as $n^{-\frac{1}{2}}$ for general integrable f. Note that the proportionality constant for this behavior is the standard target for variance reduction techniques like importance sampling, control-variates, and Rao–Blackwellization among others: for a review, see [Fis06][Gla04]. The next few sections will develop a framework that can improve on this convergence behavior, using quasi-Monte Carlo. Hence, it is complementary to these standard variance reduction techniques: it targets the behavior $n^{-\frac{1}{2}}$ and not the proportionality constant $\sigma(f)$.

2.2.3 Discrepancy: Uniformity and Integration Error

Suppose we have two different methods of numerical integration, which use the same number of points n, but the points are placed differently. We do not know anything else about the way these points are used by the two methods. Is there something we can say about the relative errors of the two methods with only this information regarding them?

One general way to address this question is to look at the properties of the quadrature point set being used, in particular the *uniformity* of the points. The following is based on Niederreiter's development of this topic in the comprehensive [Nie78]. Before discussing this more theoretically, let us see an example to illustrate the context. Figure 2.1 shows two sets of points that might be used for integration, say by a Monte Carlo algorithm. In Fig. 2.1(a) we have a 200-point "random" sample generated using a standard pseudorandom number generator (e.g., the linear congruential generator [Fis06]). In Fig. 2.1(b) we have a 200-point

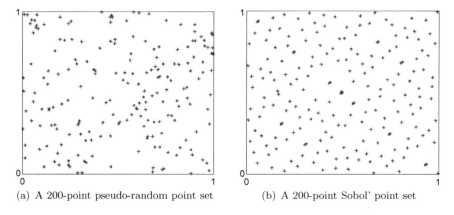

(a) A 200-point pseudo-random point set   (b) A 200-point Sobol' point set

Figure 2.1. In two dimensions, Sobol' points are more uniformly distributed than typical pseudo-random points

"deterministic" sample from the so-called Sobol' sequence. It is immediately clear that the random sample is less *uniform* than the Sobol' sample. In other words there is more *discrepancy* in the way the random points are laid out from one region to the other, as compared to the Sobol' points. The uniformity, or rather the lack of it, is often measured in terms of a quantity understandably called *discrepancy*. Hence, we say that the points in Fig. 2.1(a) have high discrepancy, while those in Fig. 2.1(b) have low-discrepancy.

The uniformity of the point set is important because we are integrating over the entire domain C^s in (2.17), and the error will tend to 0 with increasing n only if the points are drawn from a uniform distribution over the entire unit cube. Hence, at least asymptotically the points should tend towards perfect uniform distribution over C^s. For a theoretical treatment of this intuitive explanation and a comprehensive discussion on uniformity please see [KN74]. The question is that if a point set achieves better uniformity (lower discrepancy) with some fixed finite n, is the corresponding integration estimate more accurate? We now review some theoretical results that try to address this question and suggest practical implications for Monte Carlo.

There can be several definitions for discrepancy [MC94][Hic98]. The one immediately relevant to our discussion is the L_∞ *star discrepancy*, or simply the star discrepancy, which we now define. The reader may use Fig. 2.2 as a reference illustration for the following. Let us say that we have n points $\{\mathbf{x}_i : \mathbf{x}_i \in C^s\}_{i=1}^n$ in our quadrature (Monte Carlo) point set. For some hyperrectangle $J \subseteq C^s$, let $Vol(J)$ be the volume of J and let $I_J(\mathbf{x})$ be the characteristic function (2.12) for J. Define the n_J as the

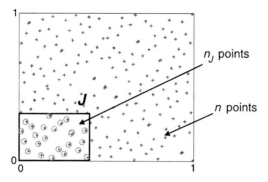

Figure 2.2. Illustration for the definition of discrepancy: n_J is the number of points inside any hyperrectangle J within the unit cube C^s

number of points lying inside J

$$n_J = \sum_{i=1}^{n} I_J(\mathbf{x}_i). \tag{2.23}$$

Then, we define the discrepancy as

$$D_n = \sup_{J \subseteq C^s} \left| \frac{n_J}{n} - Vol(J) \right|. \tag{2.24}$$

Hence, it is the maximum difference between the exact volume of any J and the estimate $(\frac{n_J}{n})$ of its volume using the points lying inside J. If we only look at hyperrectangles with one corner at the origin, $J = [\mathbf{0}, \mathbf{a})$ where $\mathbf{a} \in C^s$, then we get the star discrepancy

$$D_n^* = \sup_{\mathbf{a} \in C^s} \left| \frac{n_{[\mathbf{0},\mathbf{a})}}{n} - Vol([\mathbf{0}, \mathbf{a})) \right|. \tag{2.25}$$

The following result by Koksma in one dimension and by Hlawka in multiple dimensions provides a partial, but useful answer to our question from the beginning of this section.

THEOREM 2.4 (Koksma–Hlawka [Hla61][Nie78]). *If function f has bounded variation in the sense of Hardy and Krause, then the Monte Carlo error is bounded as follows.*

$$\epsilon(f) = |Q - Q_n| \leq V(f) D_n^* \tag{2.26}$$

where $V(f)$ is the variation of f in the sense of Hardy and Krause, and D_n^* is the star discrepancy of the point set.

$V(f)$ is a measure of the total variation of the function over the unit cube. For a smooth function in one dimension

$$V(f) = \int_0^1 |df| \qquad (2.27)$$

which is just the integral of the absolute value of the gradient of f. Hence, the more the function changes over the interval, higher is the value of $V(f)$. This can be generalized to multiple dimensions for non-smooth functions, in the sense of Hardy and Krause. The definition of this variation is not relevant for our discussion, and an intuitive understanding is sufficient. The definition is presented at the end of this section to avoid distraction.

Inequality (2.26) provides us an upper bound on the integration error for Monte Carlo using any given point set. It is particularly attractive because it separates out the two influences on the error: the properties of the function f, and the properties of the point set. Hence, it suggests that the error might be reduced if we used points with lower discrepancy. For a random sequence of points uniformly distributed over C^s, it has been shown that [Kie61]

$$D_n^* = O\left(\left[\frac{\log \log n}{n}\right]^{\frac{1}{2}}\right) \qquad (2.28)$$

with probability 1. Combining this with the deterministic bound in (2.26) we see a good match with the $n^{-0.5}$ convergence of the probabilistic error bound (2.21) for standard Monte Carlo. Taking the suggestion of Theorem 2.4 we ask if there are point sequences with lower discrepancy, and does it help to replace random sampling with these sequences?

Similar to the star discrepancy, the L_2 star discrepancy is defined as

$$T_n^* = \left[\int_{C^s} \left(\frac{n_{[\mathbf{0},\mathbf{a})}}{n} - Vol([\mathbf{0},\mathbf{a}))\right)^2 d\mathbf{a}\right]^{\frac{1}{2}}. \qquad (2.29)$$

It is known that [Nie78]

$$D_n^* \geq T_n^*. \qquad (2.30)$$

Roth [Rot80] proved a lower bound for the L_2 star discrepancy of any set of n points in C^s, which then also applies to the star discrepancy

$$D_n^* \geq T_n^* > c_s \frac{(\log n)^{\frac{s-1}{2}}}{n} \qquad (2.31)$$

where c_s depends only on s. For the first n points of an infinite sequence, it is modified [KN74] to

$$D_n^* > c_s' \frac{(\log n)^{\frac{s}{2}}}{n}. \qquad (2.32)$$

We take this opportunity to clarify the difference between a set of n points and a sequence. The former is of finite size, while latter extends to infinite size. Hence, if the required sample size n is known before hand, we can better tailor the n point locations, as compared to when the required n is not known in advance and the point generation scheme must be able to keep generating points incrementally. The bound (2.31) applies to the former, while (2.32) applies to the latter. There is a widely believed conjecture in the theory of uniform distributions that says that a tighter bound exists with the exponent $\frac{s-1}{2}$ replaced by $s-1$ in (2.31) and $\frac{s}{2}$ replaced by s in (2.32). This has been proved only for $s \leq 2$ in the first case and for $s = 1$ in the second, but it is the best seen behavior yet for arbitrary s. Hence, any such sequence, for which

$$D_n^* = O\left(\frac{(\log n)^s}{n}\right) \tag{2.33}$$

is called a *low-discrepancy sequence* (LDS) or a *quasi-random sequence*. We will use the former term in this thesis. Halton, in [Hal60b], showed the existence of infinite deterministic sequences in any dimension s which satisfy (2.33), and provided a construction for one such sequence, commonly referred to as the Halton sequence. Hence, for large values of n, we can achieve n^{-1} convergence of the discrepancy, as compared to only $n^{-0.5}$ for random sequences. The points shown in Fig. 2.1(b) are from one such LDS, discovered by Sobol' [Sob67]. We can clearly see the lower discrepancy as compared to the pseudorandom points in Fig. 2.1(a).

Other definitions and generalization for discrepancy have been proposed [Nie78][MC94][Hic98][Wo91]. In many cases corresponding results similar to the Koksma–Hlawka inequality, often for some special class of functions, have been also provided. For example, [Wo91] provides an estimate for the *average* error over a class of functions following the Brownian sheet measure – a generalization of Brownian motion to s dimensions – using the L_2 star discrepancy T_n^*.

2.2.3.1 Variation in the Sense of Hardy and Krause

Here we define the variation of a function f over C^s in the sense of Hardy and Krause, as given in [Nie78]. For any interval (hyperrectangle) in C^s, $J = [a_1^{(1)}, a_2^{(1)}] \times \cdots \times [a_1^{(s)}, a_2^{(s)}] \subseteq C^s$, define

$$\delta(f; J) = \sum_{e_1=1}^{2} \cdots \sum_{e_s=1}^{2} (-1)^{e_1+\cdots+e_s} f(a_{e_1}^{(1)}, \ldots, a_{e_s}^{(s)}). \tag{2.34}$$

Here we add up the function value at all "even" corners of J ($e_1 + \cdots + e_s$ even) and subtract out the function values at all the "odd" corners

($e_1 + \cdots + e_s$ odd). Now define any grid over C^s with any number of slices along each dimension. Each slice along any dimension can be of any arbitrary, non-trivial width, but with no overlap between slices. The set of all single cells in the grid is a *partition* $\mathcal{P}$ of C^s. Now define

$$V^{(s)}(f) = \sup_{\mathcal{P}} \sum_{J \in \mathcal{P}} |\delta(f;J)| \qquad (2.35)$$

where the supremum is over all possible partitions of C^s. This is the variation in the sense of Vitali. Let $\mathbf{u} = \{i_1, \ldots, i_k\}$ be a subset of the dimensions, such that $1 \leq k \leq s$ and $1 \leq i_1 < \cdots < i_k \leq s$. Define $C^s_{\mathbf{u}} = \{\mathbf{a} \in C^s : a_i = 1 \text{ for } a_i \notin \mathbf{u}\}$ as the subset of C^s with all coordinates not in $\mathbf{u}$ set to 1. With f restricted to $C^s_{\mathbf{u}}$, define $V^{(k)}(f; \mathbf{u})$ as the variation in the sense of Vitali over $C^s_{\mathbf{u}}$, where $k = \text{card}(\mathbf{u}) = |\mathbf{u}|$. Then, we can define $V(f)$ in the sense of Hardy and Krause as

$$V(f) = \sum_{k=1}^{s} \sum_{\{\mathbf{u}:|\mathbf{u}|=k\}} V^{(k)}(f; \mathbf{u}). \qquad (2.36)$$

If $V(f)$ is finite, then f has bounded variation in the sense of Hardy and Krause. If f is sufficiently smooth (has finite partial derivatives of sufficient order), then we can use partial derivatives instead of the finite sums and differences in (2.35), as shown in [MC94].

2.3 Low-Discrepancy Sequences

Quasi-Monte Carlo is Monte Carlo performed with points from a deterministic low-discrepancy sequence (Sect. 2.2.3). There two main classes of LDS:

1) (t,s)-sequences, and

2) integration lattices.

(t,s)-sequences have enjoyed more popularity and research than integration lattices, one reason being that it is more difficult to extend lattices to infinite sequences. In this thesis, we focus on (t,s)-sequences for these reasons. The interested reader is referred to [HHLL00][FW94][HW81] for details on integration lattices.

2.3.1 (t,m,s)-Nets and (t,s)-Sequences in Base b

In this section we present a definition of (t,m,s)-nets and (t,s)-sequences in based b, following the development in [Nie87]. As a preview, we note that a (t,m,s)-net in base b is a *fixed* set of exactly b^m points, where b is the base we will work in (e.g., 2 if binary), and m determines the size of this *finite* point set. Also, s is the number of dimensions and t is a measure of the quality of the sequence in terms of uniformity – smaller t will imply better uniformity for fixed m, s and b. These interpretations extend also to the case of (t,s)-sequences in base b, which are fixed *infinite* sequences of points in s dimensions, which are composed of an infinite number of (t,m,s)-nets in a particular manner. Smaller t still implies better uniformity, for fixed s and b. Now, we proceed towards a concrete definition.

A **b-ary box** is an interval of C^s of the form

$$J = \prod_{i=1}^{s} \left[\frac{a_i}{b^{d_i}}, \frac{a_i+1}{b^{d_i}} \right) \tag{2.37}$$

for integers $d_i \geq 0$ and $0 \leq a_i < b^{d_i}$. Hence, if we create a grid over C^s with b^{d_i} slices of equal width along dimension i, then each cell of the grid is a b-ary box. If any $d_i = 0$, then their is no slice along dimension i. Given integers $b \geq 2$ and $0 \leq t \leq m$ we can define a **(t,m,s)-net** in base b as a point set consisting of b^m points, such that $n_J = b^t$ for every b-ary box with volume $Vol(J) = b^{t-m}$. We recall from (2.23) that n_J is the number of points lying inside J, as used in the definition of star discrepancy (2.25), which we reproduce here for convenience.

$$D_n^* = \sup_{\mathbf{a} \in C^s} \left| \frac{n_{[\mathbf{0},\mathbf{a})}}{n} - Vol([\mathbf{0},\mathbf{a})) \right|. \tag{2.38}$$

Figure 2.3 illustrates this idea with a $(0,3,2)$-net in base 2: $t = 0$, $m = 3$, $s = 2$ and $b = 2$. The number of points in the net is $b^m = 2^3 = 8$. All possible 2-ary box shapes, with volume $b^{t-m} = 2^{0-3} = 1/8$ are shown cornered at the origin. Stacking any of these shapes side by side with no overlap, to fill out the unit square will give us all the 2-ary boxes with volume $1/8$ for that shape. Repeating this for all four shapes will give us all possible 2-ary boxes of volume $1/8$. We can see that every such 2-ary box contains exactly 1 ($b^t = 2^0$) point. Hence, we call this a $(0,3,2)$-net in base 2, and we say that the net *balances* all 2-ary boxes with volume $1/8$. Any box J is balanced by a net with n points, if it contains exactly $n \times Vol(J)$ points; i.e., its volume can be exactly computed using the fraction of points lying in it ($Vol(J) = n_J/n$). This property of (t,m,s)-nets helps reduce the star discrepancy (2.38) by making the term in the supremum equal to zero for some choices of $\mathbf{a}$ (for the b-ary boxes with $Vol(J) = b^{\tau-m}$, where $t \geq \tau \leq m$), and by reducing the chances of a large term for any $\mathbf{a}$. We note here that any (t,m,s)-net is also a (τ,m,s)-net for every integer $\tau \geq t$. By t we will imply the smallest such value of τ. We can see that smaller values of t lead to better uniformity, since the net can then balance, or uniformly fill, smaller boxes.

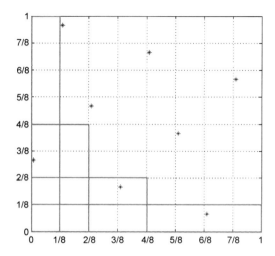

Figure 2.3. A $(0,3,2)$-net in base 2 balancing all 2-ary boxes with volume $1/8$

For base b, we write $Z_b = \{0, 1, \ldots, b-1\}$ for the set of digits in base b. For any real number $x \in [0,1]$, we can write the b-adic expansion (e.g., binary expansion for base 2) as

$$x = \sum_{j=1}^{\infty} x_j b^{-j}, \quad x_j \in Z_b, \; \forall j. \tag{2.39}$$

Truncating this at m digits (e.g., using 32 bits of computer precision), we define

$$[x]_{b,m} = \sum_{j=1}^{m} x_j b^{-j} \tag{2.40}$$

as the m-digit truncation of x. Here we make a sudden change in the notation for the coordinates of a vector: let us write any vector $\mathbf{x} \in C^s$ as $\mathbf{x} = (x^{(1)}, \ldots, x^{(s)})$. This is done for notational convenience in the following theory. Then, we can write an m-digit truncation for $\mathbf{x}$ in base b as

$$[\mathbf{x}]_{b,m} = ([x^{(1)}]_{b,m}, \ldots, [x^{(s)}]_{b,m}). \tag{2.41}$$

Let $b \geq 2$ and $t \geq 0$ be integers. Then, we define a **(t, s)-sequence** in base b as a sequence of points $\{\mathbf{x}_i : i = \{1, 2, \ldots\}\}$ in C^s, such that for all integers $k \geq 0$ and $m > t$, the set of points $\{[\mathbf{x}_i]_{b,m} : kb^m \leq i < (k+1)b^m\}$ is a (t, m, s)-net in base b. Again, we see that a sequence with smaller t for fixed s and b is preferred, since it will contain (t, m, s)-nets with smaller t. The popular constructions of Sobol' [Sob67] and Faure [Fau82] are instances of (t, s)-sequences as shown by Niederreiter [Nie87], who

Quasi-Monte Carlo

then proposed (t,s)-sequences with better properties: Niederreiter sequences [Nie88] and, subsequently Niederreiter–Xing sequences [Nie98].

In the context of exploiting low discrepancy, as defined by (2.33), we want to know the discrepancy of (t,s)-sequences. According to [Nie87], the discrepancy of a (t,s)-sequence is bounded by

$$D_n^* \leq c(t,s,b)\frac{(\log n)^s}{n} + O\left(\frac{(\log n)^{s-1}}{n}\right), \quad \forall n \geq 2. \tag{2.42}$$

This shows that any (t,s)-sequence is an LDS as defined by (2.33). $c(t,s,b)$ is independent of n, and is given by

$$c(t,s,b) = \begin{cases} \dfrac{b^t}{s}\left(\dfrac{b-1}{2\log b}\right)^s, & s=2 \text{ or } b=2,\ s=3,4 \\ \dfrac{b^t}{s!}\dfrac{b-1}{2\lfloor b/2 \rfloor}\left(\dfrac{\lfloor b/2 \rfloor}{\log b}\right)^s, & \text{otherwise} \end{cases} \tag{2.43}$$

where $\lfloor x \rfloor$ is the greater integer $\leq x$. From this, we can see that smaller values of b are preferable for given t and s, as they lower the bound on the discrepancy. The Sobol' sequences [Sob67] are in the smallest base, $b=2$, with t dependent on – and increasing with – s. It is also obvious from (2.43) that, for given b and s, smaller values of t are better, in agreement with our previous conclusion based on the definitions of (t,m,s)-nets and (t,s)-sequences. Faure's construction [Fau82] achieve the minimum vale of t, $t=0$, for b dependent on s: b is $p(s)$, the smallest prime $\geq s$.

Given this, it is natural to ask which of the two sequences is better. This is a difficult question and a clear answer is not known. In certain asymptotic terms, the discrepancy bound for the Faure sequences shows large improvement over the bound for the Sobol' sequences. The discrepancy bound constant $c(t,s,b)$ for Sobol' points takes the form

$$c_S = \frac{2^t}{s!(\log 2)^s}. \tag{2.44}$$

Sobol' [Sob67] gives the following bound for t as a function of s, for the Sobol' sequence:

$$t(s) \geq K\frac{s\log s}{\log\log s}. \tag{2.45}$$

This means that for the Sobol' sequences t increases superlinearly with increasing dimensionality s, and, thus, the constant in the discrepancy bound increases superexponentially with s. This is definitely not desirable. For the Faure sequence, the constant can be written as [Fau82]

$$c_F = \frac{1}{s!}\left(\frac{p(s)-1}{2\log p(s)}\right)^s, \tag{2.46}$$

which has the very desirable property that $\lim_{s\to\infty} C_F = 0$. However, we want to stress caution while using these *asymptotic* properties of the discrepancy *bounds* to compare the Sobol' and Faure sequences, or any other sequences. It is easily shown [MC94] that for practical values of n, the Faure bound actually increases to very large values before reducing back towards 0. We can see this by computing the maximum of the dominant term in the Faure bound. It is shown in [MC94], using basic calculus, that the maximum occurs for $n = e^s$. This shows that the dominant term in the Faure discrepancy bound increases with increasing number of points n up to $n = e^s$, whereas, it is to be expected that the actual discrepancy (uniformity) should reduce (improve) with increasing number of points. Hence, the convergence behavior of the bound sets in only after an extremely large number of points even for moderately large s.

Also, the bound in (2.42) is just that: a bound. There can be a large difference between the bound and the actual discrepancy in terms of magnitude and behavior. We cannot rely on the bound to compare the actual discrepancies of Sobol' and Faure points, as evidenced by the initial, but long increasing behavior of the Faure bound. These arguments extend for the general case of any set of (t, s)-sequences, and illustrate the difficulty in making a theory-backed choice between them. This difficulty is further worsened by the fact that the Koksma–Hlawka inequality (2.26) also only provides an upper *bound* on the integration error; and this bound may also be very loose, as shown for one class of functions in [Wo91][MC94]. [MC94] provides further insightful discussion and illustrations on this topic. Also see [Fox99], Chap. 12. We will choose the Sobol' points for our demonstrations, but based on empirical and practical considerations. However, this choice should not be taken as a definite rule for choosing Sobol' points over Faure points, since it is not backed by rigorous theoretical comparisons. We will revisit these considerations in more detail in Sect. 2.3.2.3, after we see how we can actually construct these (t, s)-sequences, along with some more examples that show better properties than both the Sobol' and Faure sequences.

2.3.2 Constructing Low-Discrepancy Sequences: The Digital Method

2.3.2.1 The Van der Corput Sequence: A Building Block

Van der Corput proposed one dimensional low-discrepancy sequences in 1935 [Van35], using b-adic expansions similar to (2.39) in some base b, where b is an integer ≥ 2. Say we are generating the n-th point, where

Quasi-Monte Carlo

n	n binary	$x_n = \psi_2(n)$ binary	x_n (fraction)
0	0	0. 0	0
1	1	0. 1	1/2
2	10	0. 01	1/4
3	11	0. 11	3/4
4	100	0. 001	1/8
5	101	0. 101	5/8
6	110	0. 011	3/8
7	111	0. 111	7/8
8	1000	0. 0001	1/16

Table 2.1. First nine points of a Van der Corput sequence in base 2

$n = 1, 2, \ldots$. Consider the b-adic expansion of $n - 1$,

$$n - 1 = \sum_{k=0}^{\infty} a_k(n) b^j = \ldots a_2 a_1 a_0, \tag{2.47}$$

where $a_k(n)$ is the k-th digit in the base b representation of $n - 1$, as represented by the last term, where we have dropped (n) to reduce notation. For finite n, only a finite number of $a_k(n)$ will be nonzero. The n-th Van der Corput point $x_n \in [0, 1)$ is then given by

$$x_n = \psi_b(n) = \sum_{k=1}^{\infty} \frac{a_{k-1}(n)}{b^k} = 0.a_0 a_1 a_2 \ldots. \tag{2.48}$$

ψ_b is the *radical inverse function* and it basically mirrors the digits about the base b radix point. A example for base 2 is shown in Table 2.1. We can see how each subsequent point is strategically placed to fill out some largest remaining gap, ensuring good uniformity over the interval $[0, 1)$.

Halton [Hal60b] provided the first method for constructing an LDS in arbitrary dimensions, by extending Hammersley's method [Ham60] of generating finite point sets with low-discrepancy. The method uses one-dimensional Van der Corput sequences [Van35] with a distinct base for each coordinate, such that the bases are relatively prime integers greater than 1. Taking the first s prime numbers is typical since smaller bases result in uniformity with fewer samples for the Van der Corput sequence. However, the Halton sequence suffers from very poor uniformity in high dimensions because of an undesirable feature of the Van der Corput sequence for large base b. We illustrate this with the example of base 10. For $n = \{1, 2, 3, 4, 5\}$, the radical inverse function ψ_{10}

gives the first 5 points as $\{0.1, 0.2, 0.3, 0.4, 0.5\}$ which are all clustered in one half of the interval $[0, 1)$. This "monotonic" filling is more pronounced for larger values of b, leading to only small parts of C^s being filled for even moderate dimensionality s. Hence, the Halton sequence is unsuited for our application where we expect to see s of the order of 10^1–10^2, and we do not dwell further on it here. For more discussion on it and its improvements, which remain inadequate, see [Gla04]. The Sobol' and Faure sequences were huge improvements over Halton's construction and enabled practical use of QMC for large dimensions. Both these methods construct each coordinate by using some generalization of the Van der Corput sequence in some base that results in a permutation of the Van der Corput sequence. In fact, as we will see, both the Sobol' and Faure constructions fall under a single general class of (t, s)-sequence constructions by Niederreiter [Nie92], called the *digital method*. The digital method uses generalizations of the Van der Corput sequence for generating the different coordinates of a sequence in s dimensions.

2.3.2.2 The Digital Method, Digital Nets and Digital Sequences

To avoid excessive technical notation, we now define digital nets and sequences only over a residue class field Z_b with b prime. The elements of Z_b are $\{0, 1, \ldots, b-1\}$. Hence, we can define any real number in base b using this field. For a more general definition over arbitrary finite commutative rings, see [Nie92]. The reduced definition is more than sufficient for our endeavors.

Digital sequence: Let $s \geq 1$ be the dimensionality, and $b \geq 2$ be a prime base. Let $\{\mathbf{C}^{(i)}\}_{i=1}^s$ be s $\infty \times \infty$ matrices over Z_b; i.e., each element of the matrices is a digit in base b. For integer $n \geq 1$ let

$$n - 1 = \sum_{k=0}^{\infty} a_k(n) b^k \qquad (2.49)$$

be the base b representation of $n - 1$. Then define a sequence $\{\mathbf{x}_n\}$ with the n-th point

$$\mathbf{x}_n = (x_n^{(1)}, \ldots, x_n^{(s)}), \qquad (2.50)$$

$$x_n^{(i)} = \sum_{k=1}^{\infty} \frac{y_k^{(i)}(n)}{b^k}, \qquad (2.51)$$

where $\{y_k^{(i)}(n)\}$ are given by

$$\begin{pmatrix} y_1^{(i)}(n) \\ y_2^{(i)}(n) \\ \vdots \end{pmatrix} = \mathbf{C}^{(i)} \cdot \begin{pmatrix} a_0(n) \\ a_1(n) \\ \vdots \end{pmatrix} \pmod{b}. \qquad (2.52)$$

Such a sequence is called a *digital sequence* over Z_b. Here it is assumed that for each n only finitely many digits $(a_k(n))$ equal $b-1$. The matrices $\mathbf{C}^{(i)}$ are called *generator matrices*. Comparing with the Van der Corput sequence in (2.48) we see that each coordinate of this digital sequence is a permuted form of the Van der Corput sequence in base b. This permutation is provided by the generator matrices. For a $\mathbf{C}^{(i)} = \mathbf{I}_\infty$ we get the original base-b Van der Corput sequence.

Digital nets: We note that the generator matrices are of size ∞ to theoretically allow an infinite digit representation. However, in practice n will always be finite, in fact, only a few thousands or millions usually, needing only a finite number of significant digits in its b-adic expansion. If we use only m digits, then only the upper left $m \times m$ submatrix of every $\mathbf{C}^{(i)}$ will be relevant, and we can generate a maximum of b^m points. With such digit truncation, we are no longer truly generating a sequence with finite matrices; we are generating finite point sets or *nets*. A point set with b^m points, so generated, is called a *digital net*. In a practical setting, say while applying this digital method for yield estimation of circuits, we will set m sufficiently high such that we never generate b^m points. Also, we will typically need to have the ability to generate points incrementally, without initial knowledge of the exact value of n we will need. In such a case, even if we are use finite generator matrices, we are effectively choosing points from the underlying infinite *sequence*. Hence, it is sufficient and also more relevant, to discuss only sequences from here on.

Digital (t,s)-sequences: A digital sequence is of use to us here only when it is a (t,s)-sequence. Niederreiter [Nie92] provides us the criteria for this requirement. Let

$$\mathbf{c}_j^{(i)} = \{c_{j1}^{(i)}, c_{j2}^{(i)}, \ldots\} \qquad (2.53)$$

be the j-th row of matrix $\mathbf{C}^{(i)}$. For integer $m > 0$ let

$$\mathbf{c}_j^{(i)}(m) = \{c_{j1}^{(i)}, c_{j2}^{(i)}, \ldots, c_{jm}^{(i)}\} \qquad (2.54)$$

be the j-th row of the upper left $m \times m$ matrix of $\mathbf{C}^{(i)}$. Define a system of vectors

$$C = \{\mathbf{c}_j^{(i)} : 1 \leq j \leq m,\ 1 \leq i \leq s\}, \qquad (2.55)$$

taking only the first m rows of the $\mathbf{C}^{(i)}$ matrices. Then for integers $0 \leq d_1, \ldots, d_s \leq m$, and $\sum_{i=1}^{s} d_i = d$, define $\sigma(C)$ as the largest d such that any subsystem $\{\mathbf{c}_j^{(i)} : 1 \leq j \leq d_i, 1 \leq i \leq s\} \subseteq C$ is linearly independent over Z_b. Now define the system of vectors

$$C(m) = \{\mathbf{c}_j^{(i)}(m) : 1 \leq j \leq m,\ 1 \leq i \leq s\}. \tag{2.56}$$

For integers $m > t \geq 0$, if $\sigma(C(m)) \geq m - t$, then the corresponding digital sequence is a (t, s)-sequence in base b. Note that we are usually interested in the smallest such t.

Some notable examples of digital (t, s)-sequences are constructions by

1) *Sobol'* [Sob67]: Sobol' constructed (t, s)-sequences in base 2 for any dimension s, with t depending on s and of order of magnitude $O(s \log s)$. This leads to a superexponential increasing behavior for the constant $c(t, s, b)$ in the leading term of the discrepancy bound (2.42) for Sobol' sequences. This was discussed in more detail in Sect. 2.3.1. The generator matrices are constructed using the coefficients of primitive polynomials over the field Z_2. A software implementation was shown in [BF88] and refined in [JK03]. We will describe the construction in detail in Sect. 2.3.3.

2) *Faure* [Fau82]: Faure constructed $(0, s)$-sequences in any prime base $b \geq s$ for any dimension s. These sequences improved the asymptotic behavior of $c(t, s, b)$ in the discrepancy bound (2.42) to $\lim_{s \to \infty} c(t, s, b) = 0$. Implication of this "improvement" and related caveats were discussed in Sect. 2.3.1. Faure used powers of the upper-triangular Pascal matrix modulo b, $\mathbf{P}_b$ to create the generator matrices:

$$\mathbf{C}^{(i)} = \mathbf{P}_b^{i-1}, \quad \text{for } i \geq 2, \qquad \mathbf{C}^{(1)} = \mathbf{I} \quad \text{(identity matrix)}. \tag{2.57}$$

The (j, k)-th element of the i-th generator matrix for $i \geq 2$ is then given by

$$c_{jk}^{(i)} = \begin{cases} {}^{k-1}C_{j-1}\, i^{(k-j)}, & j \leq k \\ 0, & j > k \end{cases}, \quad i \geq 2. \tag{2.58}$$

See [Gla04] for further details. [Fox86] presents a software implementation.

3) *Niederreiter* [Nie88]: Niederreiter generalized these previous constructions and, for the first time, showed a construction for (t, s)-sequences for all dimensions s and all bases b. For fixed b, the order of magnitude of $t = O(s \log s)$, similar to Sobol' points. However, on using different b for different s, better values of t can be obtained.

The generator matrices are constructed using coefficients of distinct monic irreducible polynomials of minimum degree. A software implementation is presented in [BFN92]. A generalization of the Halton sequence using a polynomial version of the radical inverse function (Sect. 2.3.2.1) was shown by *Tezuka* [Tez93], which is similar to the Niederreiter sequences. [NT96b] shows how the Sobol' and Faure sequences are special cases of these sequences, and obtains natural generalizations for both Sobol' and Faure sequences.

4) *Niederreiter and Xing (NX)* [XN95]: These researchers proposed the idea of using algebraic curves over finite fields (or, equivalently global function fields) to construct generator matrices, resulting in (t,s)-sequences with significantly improved theoretical quality over all previous constructions. At least four different constructions on this idea were proposed by them and are summarized in [Nie98]. For any given s, the constructions in [XN95] and [NX96] achieve the lowest values of t. The best achievable order of magnitude of t for these NX sequences is $O(s)$ for fixed b, which is significantly better than the otherwise common $O(s \log s)$. [Pir02] shows an implementation of the construction in [XN95] for dimensions 4 to 16. The construction of NX sequences require algebraic curves with certain specific properties to achieve the optimal t [Nie98]. Known examples of such algebraic curves are limited, and this limits the number of dimensions that can be constructed. Further, due to the very abstract nature of the formulation of these constructions, it is difficult for the general practitioner to implement them.

2.3.2.3 Comparing (t,s)-Sequences and Choosing One

Table 2.2, reproduced from [Nie98], shows a comparison of the t values for (t,s)-sequences constructed in base b using methods 1 [Sob67], 3 [Nie88] and 4 [NX96]. The much lower t values for the NX sequences suggest much lower discrepancy (2.43) and, consequently and potentially, much lower integration error (2.26). However, as discussed above, there are significant implementation difficulties with the NX sequences. Also, the Niederreiter sequences (method 3) do not offer significant improvement over the Sobol' or Faure sequences, for the general case. Given these reasons and the immense popularity of the Sobol' and Faure sequences among practitioners, we make a choice between the latter two options, for our experiments. Theoretical considerations do not provide a clear choice, as discussed in Sect. 2.3.1. Hence, we rely on empirical observations provided in [ABG98][Gla04] which show better performance on using Sobol' sequences. Furthermore, the fact that the Sobol' sequences

s	1) Sobol' [Sob67]	3) Nied [Nie88]	4) NX [NX96]
1	0	0	0
2	0	0	0
3	1	1	1
4	3	3	1
5	5	5	2
6	8	8	3
7	11	11	4
8	15	14	5
9	19	18	6
10	23	22	8
11	27	26	9
12	31	30	10
13	35	34	11
14	40	38	13
15	45	43	15
16	50	48	15
17	55	53	18
18	60	58	19
19	65	63	19
20	71	68	21

Table 2.2. Comparison of values of t for (t,s)-sequences in base 2, for $1 \leq s \leq 20$. The NX sequences have the lowest t values, and the best uniformity properties in terms of discrepancy (2.42). Reproduced from [Nie98]

are in base 2 allows us to exploit fast bit-level Boolean operations in the software implementation. For these reasons, we will use Sobol' sequences as our representative LDS to demonstrate the performance of QMC, and we will discuss only their construction in detail. Note that the performance of Sobol' sequences can only be improved upon by using the significantly better NX points.

2.3.3 The Sobol' Sequence

Sobol' [Sob67] gave the first construction of a (t,s)-sequence (he used the name LP_τ-sequence). Here we review the construction in the context of the digital method. We first show how the generator matrices are constructed, after which we discuss some practical issues for optimizing the uniformity of the sequences and for fast software implementation. Since each coordinate in the sequence is generated using a distinct generator matrix, let us focus on only on dimension first, and it can then be easily extended to arbitrary dimensions. Dropping the superscript for dimension, we want to compute the generator matrix **C** for a one dimensional

Quasi-Monte Carlo

Sobol' sequence. We recollect that the sequence is in base 2, hence every element of $\mathbf{C}$ is a bit: a 0 or a 1. In practice we will work with a finite number of bits for the generated values; say this is m. Then, $\mathbf{C}$ is an $m \times m$ matrix. Each *column* of this matrix can be considered as an m-bit binary expansion of some number $v_j \in [0,1)$, with an implied radix point: the uppermost element in the column is the most significant bit.

$$v_j = \sum_{k=1}^{m} \frac{c_{ki}}{2^k} = 0.c_{1i}c_{2i}\ldots c_{mi}. \tag{2.59}$$

These numbers v_j are called *direction numbers*. Using these direction numbers, we can write the digital method of (2.52) as

$$x_n = a_0(n)v_1 \oplus a_1(n)v_2 \oplus \cdots \oplus a_{m-1}(n)v_m, \quad n = 1, 2, \ldots, \tag{2.60}$$

where $\oplus$ denotes *bitwise* binary addition (modulo 2),

$$0 \oplus 0 = 0, \quad 0 \oplus 1 = 1, \quad 1 \oplus 0 = 1, \quad 1 \oplus 1 = 0, \tag{2.61}$$

which is the same as a *bitwise* XOR operation, and the a_i bits are from the binary representation of $n-1$.

Now the problem is to compute the m direction numbers. Sobol's method starts by selecting a *primitive polynomial* over $Z_2 = \{0,1\}$,

$$x^q + d_1 x^{q-1} + \cdots + d_{q-1}x + 1, \quad d_i \in \{0,1\}, \forall i. \tag{2.62}$$

This is a polynomial of degree q and coefficients d_i in $\{0,1\}$, satisfying two properties with respect to binary arithmetic (modulo 2):

- it is irreducible; i.e., it cannot be factored, and
- the smallest power p for which the polynomial divides $x^p + 1$ is $p = 2^q + 1$.

Tables listing primitive polynomials are widely available, for example in [PW72], and generation algorithms have also been suggested, as in [RB95]. We also need to choose *odd* integers $m_1, \ldots, m_q$, such that $0 < m_j < 2^j$. The polynomial (2.62) defines a recurrence relation,

$$m_j = 2d_1 m_{j-1} \oplus 2^2 d_2 m_{j-2} \oplus \cdots \oplus 2^{q-1} d_{q-1} m_{j-q+1} \oplus 2^q m_{j-q} \oplus m_{j-q},$$
$$j > q, \tag{2.63}$$

where again $\oplus$ denotes bitwise binary addition (modulo 2), or bitwise XOR. Now we can define the direction numbers as

$$v_j = \frac{m_j}{2^j}. \tag{2.64}$$

Note that dividing by 2^j is equivalent to shifting the radix point to the left j places in the binary representation of m_j. Then, we can use (2.62) to define a recurrence relation for v_j,

$$v_j = d_1 v_{j-1} \oplus c_2 v_{j-1} \oplus \cdots \oplus d_{q-1} v_{j-q+1} \oplus v_{j-q} \oplus \frac{v_{j-q}}{2^q}, \quad j > q. \tag{2.65}$$

Note that we are choosing the first q direction numbers by choosing the first q m_j values. The remaining $m - q$ direction numbers (columns of the generator matrix) can be computed using this recurrence relation.

We illustrate this procedure with an example. Consider the primitive polynomial

$$x^3 + x + 1, \tag{2.66}$$

where $q = 3$. Then the recurrence (2.64) becomes

$$v_j = v_{j-2} \oplus v_{j-3} \oplus \frac{v_{j-3}}{2^3}. \tag{2.67}$$

Suppose we initialize with $m_1 = 1, m_2 = 1, m_3 = 3$. The corresponding direction numbers are calculated by dividing m_j by 2^j, or shifting the binary radix point to the left by j places in the binary representation of m_j. Hence, in binary form

$$v_1 = m_1/2 = 0.1, \quad v_2 = m_2/2^2 = 0.01, \quad v_3 = m_3/2^3 = 0.011. \tag{2.68}$$

Also suppose that we are using $m = 5$ bits. Using the recurrence (2.67), we can compute the remaining $m - q = 5 - 3 = 2$ direction numbers:

$$v_4 = v_2 \oplus v_1 \oplus \frac{v_1}{2^3}$$
$$= 0.0100 \oplus 0.1000 \oplus 0.0001$$
$$= 0.1101,$$
$$v_5 = v_3 \oplus v_2 \oplus \frac{v_2}{2^3}$$
$$= 0.01100 \oplus 0.01000 \oplus 0.00001$$
$$= 0.00101. \tag{2.69}$$

Using the bits of these direction numbers, we can write our generator matrix as

$$\mathbf{C} = \begin{pmatrix} 1 & 0 & 0 & 1 & 0 \\ 0 & 1 & 1 & 1 & 0 \\ 0 & 0 & 1 & 0 & 1 \\ 0 & 0 & 0 & 1 & 0 \\ 0 & 0 & 0 & 0 & 1 \end{pmatrix}. \tag{2.70}$$

Note that the generator matrix is an upper diagonal matrix. This is true in general for any Sobol' generator matrix. This is because the j-th column (direction number) is generated by taking a number m_j with maximum j bits ($m_j < 2^j$). Also, every diagonal element is 1 because every m_j is odd. We can use this generator matrix in (2.52) to generate the Sobol' points. Instead, equivalently, we use (2.60) exploiting efficient bitwise binary operations.

$$x_1 = 0(0.10000) \oplus 0(0.01000) \oplus 0(0.01100) \oplus 0(0.11010) \oplus 0(0.00101)$$
$$= 0.0 = 0$$
$$x_2 = \mathbf{1}(0.10000) \oplus 0(0.01000) \oplus 0(0.01100) \oplus 0(0.11010) \oplus 0(0.00101)$$
$$= 0.1 = 1/2$$
$$x_3 = 0(0.10000) \oplus \mathbf{1}(0.01000) \oplus 0(0.01100) \oplus 0(0.11010) \oplus 0(0.00101)$$
$$= 0.01 = 1/4$$
$$x_4 = \mathbf{1}(0.10000) \oplus \mathbf{1}(0.01000) \oplus 0(0.01100) \oplus 0(0.11010) \oplus 0(0.00101)$$
$$= 0.11 = 3/4$$
$$x_5 = 0(0.10000) \oplus 0(0.01000) \oplus \mathbf{1}(0.01100) \oplus 0(0.11010) \oplus 0(0.00101)$$
$$= 0.011 = 3/8$$
$$x_6 = \mathbf{1}(0.10000) \oplus 0(0.01000) \oplus \mathbf{1}(0.01100) \oplus 0(0.11010) \oplus 0(0.00101)$$
$$= 0.111 = 7/8$$
$$x_7 = 0(0.10000) \oplus \mathbf{1}(0.01000) \oplus \mathbf{1}(0.01100) \oplus 0(0.11010) \oplus 0(0.00101)$$
$$= 0.001 = 1/8$$
$$\vdots = \vdots \tag{2.71}$$

We can see that the resulting points are permutations of the Van der Corput sequence in base 2 (Sect. 2.3.2.1). For the case of multiple dimensions ($s > 1$), each dimension gets its own distinct primitive polynomial and a corresponding set of initial m_j values. This leads to different permutations of the Van der Corput sequence in different dimensions, resulting in uniform distribution in the sampling region C^s.

The Sobol' construction takes two external inputs for each dimension: the primitive polynomial and the set of m_j values. Two natural questions that follow are:

- How do these inputs affect the properties of the resulting sequence?
- What are good choices for these inputs?

Sobol' provides us with some answers to these questions. First we look at the choice of polynomials.

2.3.3.1 Choosing Primitive Polynomials for Good Sobol' Sequences

Sobol' [Sob67] showed that under certain conditions the t parameter for a Sobol' sequence is

$$t = \sum_{i=1}^{s}(q_i - 1) = q_1 + q_2 + \cdots + q_s - d, \qquad (2.72)$$

where q_i is the degree of the primitive polynomial used for dimension i. In the general case, a Sobol' sequence might achieve a lower t value: this is an upper bound on the lowest t value, that is exact under the conditions given in [Sob67]. Since a lower value of t leads to better uniformity and a lower bound on the discrepancy (Sect. 2.3.1), this result recommends using polynomials of lowest possible degree. Hence, we sort the polynomials with nondecreasing degree and use them in the same order for increasing dimensions.

2.3.3.2 Choosing Initial Direction Numbers for Good Sobol' Sequences

Sobol' [Sob76] defines two uniformity properties for any sequences:

- Property A: An s-dimensional sequence $\{\mathbf{x}_n\}$ satisfies property A if for every $j = 0, 1, \ldots$ exactly one of the points $\{\mathbf{x}_k : j2^s \leq k < (j+1)2^s\}$ falls in each of the 2^s cubes of the form

$$\prod_{i=1}^{s}\left[\frac{a_i}{2}, \frac{a_i+1}{2}\right), \quad a_i \in \{0,1\}. \qquad (2.73)$$

In other words, every set of points $\{\mathbf{x}_k : j2^s \leq k < (j+1)2^s\}$ is a $(0, s, s)$-net in base 2. Note that some similar properties are satisfied by any (t, s)-sequence (Sect. 2.3.1), but this property strengthens the uniformity requirement.

- Property A': An s-dimensional sequence $\{\mathbf{x}_n\}$ satisfies property A' if for every $j = 0, 1, \ldots$ exactly one of the points $\{\mathbf{x}_k : j2^{2s} \leq k < (j+1)2^{2s}\}$ falls in each of the 2^{2s} cubes of the form

$$\prod_{i=1}^{s}\left[\frac{a_i}{4}, \frac{a_i+1}{4}\right), \quad a_i \in \{0,1,2,3\}. \qquad (2.74)$$

In other words, every set of points $\{\mathbf{x}_k : j2^{2s} \leq k < (j+1)2^{2s}\}$ is a $(0, 2s, s)$-net in base b. Once again, although there are similarities with the (t, m, s)-net properties of a (t, s)-sequence, this property strengthens the uniformity requirement.

Sobol' [Sob76] also provides conditions on the direction numbers to ensure these additional uniformity properties for the resulting Sobol' sequences. Denote the j-th direction number for the i-th dimension by $v_j^{(i)}$. Thus, the generator matrix $\mathbf{X}^{(i)}$ is composed from $\{v_1^{(i)}, v_2^{(i)}, \ldots\}$, where we have used the column vector interpretation of each $v_j^{(i)}$. Denote the first bit of $v_j^{(i)}$ by $v_{j,1}^{(i)}$: this is also the first element of the j-th column of $\mathbf{C}^{(i)}$ or, equivalently, the j-th element of the first row of $\mathbf{C}^{(i)}$. Then, property A holds for the generated sequence if and only if

$$\begin{vmatrix} v_{1,1}^{(1)} & v_{2,1}^{(1)} & \cdots & v_{s,1}^{(1)} \\ v_{1,1}^{(2)} & v_{2,1}^{(2)} & \cdots & v_{s,1}^{(2)} \\ \vdots & \vdots & \ddots & \vdots \\ v_{1,1}^{(s)} & v_{2,1}^{(s)} & \cdots & v_{s,1}^{(s)} \end{vmatrix} \neq 0 \bmod 2. \qquad (2.75)$$

Note that this condition is on the first s direction numbers. In practice we will use m direction numbers and for large s, m may be less than s. However, theoretically, all s direction numbers do exist from their recurrence relation (2.64). Following the notation from above, let us denote the *second* bit of $v_j^{(i)}$ by $v_{j,2}^{(i)}$: this is also the second element of the j-th column of $\mathbf{C}^{(i)}$ or, equivalently, the j-th element of the second row of $\mathbf{C}^{(i)}$. Sobol' also shows that property A' holds if and only if

$$\begin{vmatrix} v_{1,1}^{(1)} & v_{2,1}^{(1)} & \cdots & v_{2s,1}^{(1)} \\ v_{1,2}^{(1)} & v_{2,2}^{(1)} & \cdots & v_{2s,2}^{(1)} \\ \vdots & \vdots & \ddots & \vdots \\ v_{1,1}^{(s)} & v_{2,1}^{(s)} & \cdots & v_{2s,1}^{(s)} \\ v_{1,2}^{(s)} & v_{2,2}^{(s)} & \cdots & v_{2s,2}^{(s)} \end{vmatrix} \neq 0 \bmod 2. \qquad (2.76)$$

We note that property A applies to subsequences of length 2^s, whereas property A' applies to subsequences of length 2^{2s}. Hence, even for moderately large s (order of 10), property A is of more interest to us in practical settings. Bratley and Fox [BF88] provide values of m_j that satisfy property A, for up to 40 dimensions. Joe and Kuo [JK03] propose a method to compute good m_j values and further extend the list to 1,111 dimensions.

2.3.3.3 Gray Code Construction

Antanov and Saleev [AS79] show that the implementation of Sobol's construction is simplified if the binary representation $\{a_0(n), \ldots, a_{m-1}(n)\}$ of $n-1$ in (2.60) is replaced by the Gray code representation $\{g_0(n), \ldots, g_{m-1}(n)\}$ of $n-1$. They show that this does not affect the asymptotic discrepancy behavior of the sequence. The binary Gray code can be obtained from the binary representation using

$$g_{m-1} \ldots g_1 g_0 = a_{m-1} \ldots a_1 a_0 \oplus 0 a_{m-1} \ldots a_1, \qquad (2.77)$$

where a_i is the i-th significant bit in the binary representation and g_i is the corresponding i-th bit in the Gray code representation. The reason for the simplification is that the Gray code of subsequent integers $n-1$ and n differ only in one bit. Let us rewrite the Sobol' point x_n (2.60) in one dimension as

$$x_n = g_0(n) v_1 \oplus g_1(n) v_2 \oplus \cdots \oplus g_{m-1}(n) v_m \qquad (2.78)$$

using the Gray code of $n-1$. Suppose the Gray codes of $n-1$ ($\{g_i(n)\}$) and n ($\{g_i(n+1)\}$) differ in the l-th bit. Then, we can write

$$\begin{aligned} x_{n+1} &= g_0(n+1) v_1 \oplus g_1(n+1) v_2 \oplus \cdots \oplus g_{m-1}(n+1) v_m \\ &= g_0(n) v_1 \oplus g_1(n) v_2 \oplus \cdots \oplus (g_l(n) \oplus 1) v_l \oplus g_{m-1}(n) v_m \\ &= x_n \oplus v_l. \end{aligned} \qquad (2.79)$$

Hence, the points can be computed recursively, using only one bitwise XOR operation instead of m in (2.60). In the next section, we take a diversion to review Latin hypercube sampling, which is a popular Monte Carlo sampling technique that also tries to ensure good uniformity, and has been suggested for use on circuit problems [SKC99].

2.3.4 Latin Hypercube Sampling

Latin hypercube sampling (LHS), introduced in [MBC79], is a variance reduction technique applied to Monte Carlo. Recalling the asymptotic Monte Carlo variance (2.21)

$$\sigma^2_{\text{MC}} = \frac{\sigma^2(f)}{n}, \qquad (2.80)$$

LHS reduces this variance by reducing the contribution of $\sigma^2(f)$. LHS is effective for functions that can be largely separated into a sum of one dimensional functions, each one depending on only one of the input variables. We will discuss this in more concrete terms after introducing the concepts of ANOVA decomposition and effective dimension in

Quasi-Monte Carlo

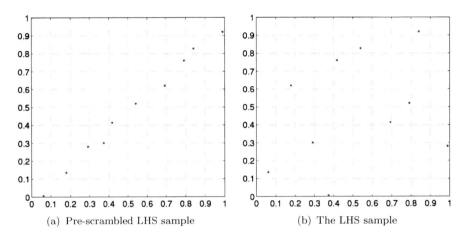

Figure 2.4. Latin hypercube sample of 10 points in 2 dimensions

Sect. 2.4.1. For now we review the construction of an LHS sample and satisfy ourselves with an intuitive, but convincing argument of its variance reduction effectiveness. We also discuss the connection between an LHS sample and (t, m, s)-nets.

2.3.4.1 Construction

Suppose we want to generate n points uniformly distributed in the s dimensional unit cube C^s. For this, a necessary condition is that the marginal distribution along each dimension should be uniform. Latin hypercube sampling tries to ensure good uniformity along each dimension as follows. Divide each dimension into n equal slices, or *strata*, forming a grid of n^s equal cells. For each stratum $j = \{1, \ldots, n\}$ in dimension i, independently draw a uniformly distributed random value $y_j^{(i)}$ within the stratum. Generate n such values independently for each dimension:

$$y_j^{(i)} = \frac{j - 1 + U_j^{(i)}}{n}, \quad i = 1, \ldots, s, \ j = 1, \ldots, n, \qquad (2.81)$$

where $U_j^{(i)}$ are uniformly distributed, independent random variables over $[0, 1)$. This gives us n random values for each coordinate i, resulting in n points in C^s. However, the coordinate values for point j lie within the same stratum j along each dimension, resulting in points that are arranged in the diagonal cells. An example is shown in Fig. 2.4(a): the dotted lines show the strata along each dimension. This is definitely not uniformly distributed in the sense we desire. To achieve this uniform distribution, we randomly rearrange the strata along each dimension as follows. For dimension i fix a permutation $\pi_i : \{1,\ldots,n\} \to \{1,\ldots,n\}$: this essentially "scrambles" the slices, and hence the coordinate values, along dimension i. s permutations $\pi_1,\ldots,\pi_s$, one for each dimension, are randomly drawn from the set of $n!$ such permutations. Denote by $\pi_i(j)$ the permuted value of j: the j-th stratum along dimension i is scrambled to location $\pi_i(j)$. This scrambling of the strata along each dimension also causes a scrambling of the sampled coordinate values as

$$x_j^{(i)} = y_{\pi_i(j)}^{(i)} = \frac{\pi_i(j) - 1 + U_j^{(i)}}{n}, \quad i=1,\ldots,s,\ j=1,\ldots,n, \qquad (2.82)$$

with an independent scrambling scheme for each dimension. Now, we can compose the points in the LHS as

$$\mathbf{x}_j = \{x_j^{(1)}, x_j^{(2)}, \ldots, x_j^{(s)}\}, \quad j=1,\ldots,n. \qquad (2.83)$$

Figure 2.4(b) shows the resulting set of points after scrambling the points from Fig. 2.4(a). Note that the coordinate values of the points are not changed by this scrambling, only the relative ordering is. Hence, the points have very good uniformity along each dimension.

2.3.4.2 Variance (and Integration Error) Reduction

LHS is a special case of *stratified sampling* [Fis06][Gla04] – a popular, general method for variance reduction – because it stratifies each dimension. This stratification tries to ensure that the sample points are well spread out over the unit cube and there is not much variation in the way the integrand f is sampled if we generate different LHS samples with the same number of points. As a result there is also less variation in the integral estimate Q_n from one LHS run to another. Compare this with standard Monte Carlo, where due to lack of any such stratification, there is some chance that in two different runs the points will be clustered together in two different parts of the unit cube. This can result in large variation in the way f is sampled, and in the estimate Q_n. Hence, we often see a decrease in the variance of Q_n, and hence, the integration error, on using Latin hypercube sampling instead of standard Monte Carlo. McKay et al. [MBC79] derive the following result for the asymptotic variance of LHS.

$$\sigma_{\text{LHS}}^2 = \sigma_{\text{MC}}^2 + \frac{n}{n-1}\text{Cov}(\mu_1,\mu_2), \qquad (2.84)$$

where μ_1,μ_2 are the mean values of f over any two cells in the grid resulting from the stratification, and Cov is the covariance, computed by taking the expectation over all possible pairs of cells. The paper

Quasi-Monte Carlo

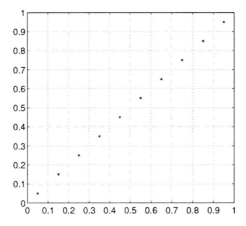

Figure 2.5. $(0,1,2)$-net in base 10: pre-scrambled non-perturbation version of the LHS sample in Fig. 2.4(b)

also shows conditions when the second component of the variance can be negative, resulting in a variance reduction: when f is monotonic in each of its inputs. We review a different mathematical treatment of the variance reduction process of LHS, along with more general conditions on f that lead to efficient variance reduction, in Sect. 2.4.2.

2.3.4.3 LHS Sample Is a Scrambled (t, m, s)-Net

It is obvious from the construction that the sample size n is required in advance to generate an LHS sample, and arbitrary additions to the sample are not possible. One LHS run, then, generates a fixed set of points, also called a net as in our discussion of (t, m, s)-nets in Sect. 2.3.1. One popular construction of LHS samples replaces $U_j^{(i)}$ by $1/2$ in (2.82), placing every point at the exact center of the cell containing it. This improves the uniformity of the sample along each dimension, but increases the bias in the integral estimate: as the number of points is increased, Q_n does not tend exactly to Q. However, this error is often relatively small compared to the variance for practical sample sizes. The *pre-scrambled* version of such an LHS sample, with coordinate values given by

$$y_j^{(i)} = \frac{j - 1 + 0.5}{n}, \quad i = 1, \ldots, s, \ j = 1, \ldots, n, \qquad (2.85)$$

is a $(0, 1, s)$-net in base n. This is because it is a set of n points that balances every n-ary box (refer (2.37)) of volume $n^{0-1} = 1/n$, as required by the definition of a $(0, 1, s)$-net in base n. The resulting LHS sample is a *scrambled* $(0, 1, s)$-net in base n, which has the same uniformity properties as the pre-scrambled version. The complete construction, us-

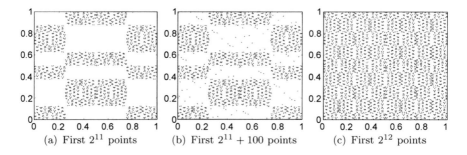

Figure 2.6. Dimensions 38 and 39 of a 40-dimensional Sobol' sequence showing undesirable patterns

ing $U_j^{(i)}$ as in (2.82) results in a scrambled $(0,1,s)$-net in base n with small random perturbations to the coordinates. The non-perturbed prescrambled $(0,1,2)$-net in base 10 is shown in Fig. 2.5 for our two dimensional LHS example. We will further discuss scrambled nets and sequences in Sect. 2.5.3 in the context of extracting variance estimates for quasi-Monte Carlo. We will also revisit LHS in Sect. 2.4.2 where we compare it to Sobol' points.

2.4 Quasi-Monte Carlo in High Dimensions

A necessary, but not sufficient, condition for uniformity in all dimensions is uniformity in low dimension projections. Suppose we generate an $s = 40$ dimensional Sobol' point set. Now pick any two coordinates and plot the values of those coordinates. If we do not see a uniform distribution in the results space $[0,1)^2$, then the point set is not uniform in 40 dimensions. Figure 2.6(a) plots coordinates 38 and 39, corresponding to primitive polynomials $x^8 + x^4 + x^3 + x^2 + 1$ and $x^8 + x^6 + x^5 + x^4 + 1$, respectively, for the first 2,048 Sobol' points. It is obvious that the projection is not uniform, and hence, the 40-dimensional Sobol' point set is not uniform. As we increase the number of points, the gaps get filled out and we achieve good uniformity in the projection, and ultimately in all dimensions.

We can see why this happens if we refer to the definition of a (t,s)-sequence in base b given in Sect. 2.3.1. Given $m > t$, we need at least b^{t+1} points for the (t,m,s)-net property to manifest. The minimum uniformity criterion for all dimensions in this context is to balance (equally fill) all b-ary boxes (2.37) resulting from the minimum number of non-trivial slices along each dimensions ($d_i > 0, \forall i$). The minimum number of slices is b, resulting in b^s boxes, each of volume b^{-s}. To balance these boxes with points from a (t,s)-sequence, a minimum of b^{t+s} points are

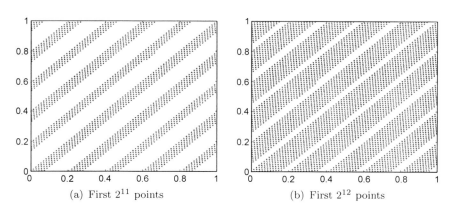

Figure 2.7. Dimensions 1 and 30 of a 70-dimensional Faure sequence showing undesirable patterns

required, such that every box has b^t points. The asymptotic discrepancy rate (2.42) therefore starts only from $n = b^{t+s}$ points. From Table 2.2 of t values, we can see that the required n can become astronomical for large s. For our Sobol' example, we use 8-th order polynomials for both dimensions 38 and 39. From the t value bound of (2.72), this 2 dimensional projection has a t value of $8 + 8 - 2 = 14$ or less. Hence, we should expect to see good uniformity with 2^{14} points. If we add the next 100 points, we see that the gaps start to get filled in Fig. 2.6(b), and we actually reach good uniformity with 2^{12} points (Fig. 2.6(c)), indicating that the t value for this two dimensional projection might be less than 14. We see similar patterns with other low discrepancy sequences too; Fig. 2.7 shows an example for a sample from a Faure sequence [Fau82].

Lack of uniformity in high dimensions is also evidenced by the asymptotic discrepancy behavior

$$D_n^* = O\left(\frac{(\log n)^s}{n}\right). \qquad (2.86)$$

For large s, the numerator dominates for practical values of n. An extremely large number of samples is required for the denominator to dominate and for the rate to improve to n^{-1}. Despite this impractically large values of n, QMC has been found to work well for several high-dimensional problems in finance [NT96b][PT95][Gla04] with realistic sample sizes. The reason for this is that low dimensional projections of QMC points can have small discrepancy, and this can exploit dominant low dimensional structure of the integrand. For example, the two dimensional projection of Fig. 2.6 has a t value of ≤ 14, while the original 40 dimensional sequence has a much larger t value (≤ 193). We next review some theoretical concepts that provide us guidelines for exploiting this feature.

2.4.1 Effective Dimension of the Integrand

Assume function $f \in L_2(C^s)$ is square integrable over C^s. Let $\mathbf{u} \subseteq \{1,\ldots,s\}$ denote any subset of the input dimensions of f. We use $|\mathbf{u}|$ for its cardinality and $-\mathbf{u}$ for its complementary set $\{1,\ldots,s\} - \mathbf{u}$. Then, for any point $\mathbf{x} = \{x_1,\ldots,x_s\} \in C^s$, $\mathbf{x_u} = \{x_i : i \in \mathbf{u}\}$ is the vector of the coordinates of $\mathbf{x}$ belonging to $\mathbf{u}$. $C^{\mathbf{u}}$ is the unit cube in the dimensions belonging to $\mathbf{u}$. We can write f as the sum of 2^s "simpler" functions using its *analysis of variance* (ANOVA) decomposition as

$$f(\mathbf{x}) = \sum_{\mathbf{u} \subseteq \{1,\ldots,s\}} f_{\mathbf{u}}(\mathbf{x}), \qquad (2.87)$$

where each function $f_{\mathbf{u}}$ depends only on $\mathbf{x_u}$, excluding the effect of the proper subsets of $\mathbf{x_u}$. We note that the integral of f over $-\mathbf{u}$ is a function of only $\mathbf{x_u}$. For example, if $s = 2$, $f = x_1^2 + x_2^3$, $\mathbf{u} = \{1\}$ and $-\mathbf{u} = \{2\}$,

$$\int_{C^{\{2\}}} (x_1^2 + x_2^3) dx_2 = x_1^2 + 1/4. \qquad (2.88)$$

Hence, the ANOVA terms are defined as

$$f_{\mathbf{u}}(\mathbf{x}) = \int_{C^{-\mathbf{u}}} \left(f(\mathbf{x}) - \sum_{\mathbf{v} \subset \mathbf{u}} f_{\mathbf{v}}(\mathbf{x}) \right) d\mathbf{x}_{-\mathbf{u}} \qquad (2.89)$$

$$= \int_{C^{-\mathbf{u}}} f(\mathbf{x}) d\mathbf{x}_{-\mathbf{u}} - \sum_{\mathbf{v} \subset \mathbf{u}} f_{\mathbf{v}}(\mathbf{x}). \qquad (2.90)$$

To compute the function $f_{\mathbf{u}}$ we subtract out the effect of all the proper subsets of $\mathbf{u}$ and then average over the dimensions that are not in $\mathbf{u}$. Note that for the empty set, $f_{\emptyset}(\mathbf{x}) = \int_{C^s} f(\mathbf{x}) d\mathbf{x} = Q$ is a constant equal to the integral of f. These ANOVA terms enjoy the following properties.

- $\int_0^1 f_{\mathbf{u}} dx_j = 0$ for any $j \in \mathbf{u}$.
- The ANOVA decomposition is orthogonal: $\int_{C^s} f_{\mathbf{u}} f_{\mathbf{v}} d\mathbf{x} = 0$ if $\mathbf{u} \neq \mathbf{v}$.
- If $\sigma^2 = \int_{C^s} (f(\mathbf{x}) - Q)^2 d\mathbf{x}$ is the variance of f, then $\sigma^2 = \sum_{|\mathbf{u}|>0} \sigma_{\mathbf{u}}^2$, where $\sigma_{\mathbf{u}}^2 = \int_{C^s} f_{\mathbf{u}}(\mathbf{x})^2 d\mathbf{x}$ is the variance of $f_{\mathbf{u}}$. Note that $\sigma_{\emptyset} = 0$ since $f_{\emptyset}(\mathbf{x})$ is constant.

We can use the variance contribution $\sigma_{\mathbf{u}}^2$ of any $f_{\mathbf{u}}$ to measure the relative importance of $f_{\mathbf{u}}$. In fact, normalized variances $\sigma_{\mathbf{u}}^2/\sigma^2$ are used as such measures and are called *global sensitivity indices* in [SK05]. These are similar, but more general, in concept to the relative global sensitivity metric proposed in Sect. 1.6.3.1 of this thesis.

We now make some observations regarding f using its ANOVA decomposition. Let $g_t(\mathbf{x}) = \sum_{|\mathbf{u}|=t} f_\mathbf{u}(\mathbf{x})$ for $0 \leq t \leq s$. Then g_t captures that part of f that depends on t dimensional inputs, or, in other words, the part that is exactly t dimensional. Consequently, $\sum_{i=1}^{t} g_i(\mathbf{x})$ is that part of f that is at most t dimensional. From the orthogonality of the ANOVA terms, it follows that the variance of g_t is $\sigma^2(g_t) = \sum_{|\mathbf{u}|=t} \sigma_\mathbf{u}^2$. If, for some f, $\sigma^2(g_1) \geq 0.99\sigma^2$, then most (99%) of the variance of f is contributed by one dimensional ANOVA terms, and we say that f is effectively one dimensional in its inputs. Similarly, if $\sigma^2(g_1) + \sigma^2(g_2) + \sigma^2(g_3) \geq 0.99\sigma^2$, then f is effectively three dimensional in its inputs. Caflish et al. [CMO97] formalize two measures of the *effective dimension* of any function f.

Superposition sense: *The effective dimension of f with variance σ^2, in the superposition sense, is the smallest integer s_S such that*

$$\sum_{0<|\mathbf{u}|\leq s_S} \sigma_\mathbf{u}^2 \geq 0.99\sigma^2. \quad (2.91)$$

Truncation sense: *The effective dimension of f with variance σ^2, in the truncation sense, is the smallest integer s_T such that*

$$\sum_{\mathbf{u} \subseteq \{1,\ldots,s_T\}} \sigma_\mathbf{u}^2 \geq 0.99\sigma^2. \quad (2.92)$$

From these definitions, we can see that $s_S \leq s_T$. Wang and Fang [WF03] extend Sobol's method of computing ANOVA variances [SK05], to compute effective dimension. The proposed technique uses extensive Monte Carlo runs to approximate the variances of the ANOVA terms. The threshold of 0.99 in the definitions is arbitrary; other values may be preferable in different setting. s_S is an indicator of whether only low dimensional interactions dominate the variance in f, while s_T is the number of leading dimensions, given an ordering, that account for most of the variance in f. For example, if $f = x_1 + x_2 + x_4$, only one dimensional "interactions" can be added up to explain f. Hence, $s_S = 1$. However, the four leading dimensions $\{x_1, \ldots, x_4\}$ are needed to explain at least 99% of the variance of f. Hence, $s_T = 4$. Note that if we reorder the dimensions by swapping x_3 and x_4, then only the leading 3 dimensions are needed and s_T is now reduced to 3. We will rely on such reorderings to make QMC effective even in high dimensions for our circuit analysis problems. We note here that, typically, circuit performance metrics are significantly affected only by some small subset of the parameters in the circuit. This claim is supported by results from experiments on the SiLVR model proposed in Chap. 1 of this thesis. These results are presented in Sect. 1.7. Hence, in many cases, the integrand for a statistical circuit analysis problem (e.g., Sect. 2.2.1.2) has a low effective dimension, at least in the truncation sense, assuming a proper ordering of the statistical parameters. The following result from [CMO97] provides hints as to how we can exploit this feature of our integrands.

For an n-point sample from an LDS, let $D^*_{n,\mathbf{u}}$ be the star discrepancy of the $|\mathbf{u}|$ dimensional points obtained by selecting only the coordinates in $\mathbf{u}$. For example the discrepancy of the two dimensional projection of the Sobol' points on $\mathbf{u} = \{38, 39\}$ is denoted by $D^*_{n,\{38,39\}}$. Denote the integration error of an n-point quadrature on any function f by

$$e_n(f) = \left| \int_{C^s} f(\mathbf{x}) d\mathbf{x} - \frac{1}{n} \sum_{i=1}^{n} f(\mathbf{x}_i) \right|, \qquad (2.93)$$

where $\{\mathbf{x}_i\}_{i=1}^n$ are the quadrature points: the sample points for Monte Carlo or QMC. We can write this error using the ANOVA terms of f as

$$e_n(f) = \left| \int_{C^s} \sum_{\mathbf{u} \subseteq \{1,\ldots,s\}} f_{\mathbf{u}}(\mathbf{x}) d\mathbf{x} - \frac{1}{n} \sum_{i=1}^{n} \sum_{\mathbf{u} \subseteq \{1,\ldots,s\}} f_{\mathbf{u}}(\mathbf{x}_i) d\mathbf{x} \right|$$

$$= \left| \sum_{\mathbf{u} \subseteq \{1,\ldots,s\} : |\mathbf{u}|>0} \left[\int_{C^{\mathbf{u}}} f_{\mathbf{u}}(\mathbf{x}_{\mathbf{u}}) d\mathbf{x}_{\mathbf{u}} - \frac{1}{n} \sum_{i=1}^{n} f_{\mathbf{u}}((\mathbf{x}_i)_{\mathbf{u}}) \right] \right|$$

(using $f_\emptyset(\mathbf{x}) = $ constant Q)

$$\leq \sum_{\mathbf{u} \subseteq \{1,\ldots,s\} : |\mathbf{u}|>0} \left| \int_{C^{\mathbf{u}}} f_{\mathbf{u}}(\mathbf{x}_{\mathbf{u}}) d\mathbf{x}_{\mathbf{u}} - \frac{1}{n} \sum_{i=1}^{n} f_{\mathbf{u}}((\mathbf{x}_i)_{\mathbf{u}}) \right|$$

(using the triangle inequality)

$$= \sum_{\mathbf{u} \subseteq \{1,\ldots,s\} : |\mathbf{u}|>0} e_n(f_{\mathbf{u}}). \qquad (2.94)$$

Then, using the Koksma–Hlawka inequality (2.26) for each ANOVA term, we can write

$$e_n(f) \leq \sum_{\mathbf{u} \subseteq \{1,\ldots,s\} : |\mathbf{u}|>0} e_n(f_{\mathbf{u}}) \leq \sum_{\mathbf{u} \subseteq \{1,\ldots,s\} : |\mathbf{u}|>0} V_{\mathbf{u}}(f_{\mathbf{u}}) D^*_{n,\mathbf{u}}, \qquad (2.95)$$

where $V_{\mathbf{u}}(f_{\mathbf{u}})$ is the variation of $f_{\mathbf{u}}$ taken as a function over $C^{\mathbf{u}}$. Different versions of this relation are given in [WS07]. Also see [Hic98]. If, for the subsets $\mathbf{u}$ that show large variance $\sigma_\mathbf{u}$ (related to $V_\mathbf{u}(f_\mathbf{u})$), the discrepancy $D^*_{n,\mathbf{u}}$ of the projection of the LDS onto $\mathbf{u}$ is small, then all the terms in the error bound are small, leading to a small error bound. This suggests two possible ways of achieving low integration error in high dimensions with QMC points that are not very uniform in high dimensions. In both these ways we exploit any low effective dimension properties of the integrand, for example, the integrand for circuit yield analysis.

1) If the effective dimension of f in the superposition sense, s_S, is small, it may be possible achieve very low integration error with an LDS that has good uniformity in low dimensional projections. From the discussion at the beginning of this Sect. 2.4, we know that QMC points can achieve good uniformity in low dimensional projections. This may not be true for some combinations of dimensions, as shown in Fig. 2.6, but on average the low dimension projections can show better uniformity than pseudorandom points even for realistic sample sizes [WF03][WS07]. In other words, for small $|\mathbf{u}|$, $D^*_{n,\mathbf{u}}$ is often small. If s_S is small then the high variance ANOVA terms are functions on subsets $\mathbf{u}$ with small $|\mathbf{u}|$. Hence, from (2.95), this can lead to low integration errors, even if the overall discrepancy (2.86) of the LDS is large.

2) If the effective dimension of f in the truncation sense, s_T, is small, we can lower the error bound with an LDS that is uniform in the first s_T dimensions, even if the higher dimensions are not sampled uniformly. Typically, the initial dimensions are sampled more uniformly than the higher dimensions by samples from an LDS. Hence, $D^*_{n,\mathbf{u}}$ can be small for $\mathbf{u} = \{1, \ldots, s_T\}$ if s_T is not too large. We can see this for the case of Sobol' points, where the early dimensions are generated using the lowest degree primitive polynomials. Hence, using (2.72) the t value of the sequence containing only the first s_T dimensions (e.g., dimensions 1–10) will be lower than the t value of any sequence containing any higher s_T dimensions (e.g., dimensions 91–100). For circuit yield problems, low s_T can be achieved for the integrand by arranging the statistical parameters in decreasing order of their impact on the relevant circuit performance, assuming that the total number of important parameters is not large.

Researchers in finance use linear transformations such as *Brownian bridge* (BB) and *principal components analysis* (PCA) on the input variables to reduce the truncation dimension: most of the variance in the resulting joint probability distribution of the transformed inputs is concentrated in the early dimensions. Extensive experiments showing the advantage of exploiting reduced effective dimension can be found in [MC96][CMO97][ABG98][WF03][Owe03b]. For example, [WF03] shows that for the problem of pricing an Asian option, the truncation dimension can be reduced from 53 to 2 using PCA for a 64 dimensional problem, resulting in large reductions in error and much improved convergence. PCA is widely used in the electronic design automation community for reducing the number of statistical parameters to a few dominant ones that explain most of their variance [Ism93][CS05][LLPS05][LLP04]. In our proposed framework, we start from the result of any such PCA, and must further reduce the truncation dimension without the option of using PCA. Hence, we skip a detailed explanation. For details regarding PCA and BB, please refer to [Gla04].

2.4.2 Why Is Quasi-Monte Carlo (Sobol' Points) Better Than Latin Hypercube Sampling?

We saw in Sect. 2.3.4 that an LHS sample is essentially a scrambled (t,m,s)-net, specifically a $(0,1,s)$-net in base n, where n is the sample size. Hence, it is natural to ask if we gain any improvement by moving to a more general QMC approach, say using Sobol' sequences? If yes, then what are the reasons for such improvement? With the knowledge of ANOVA decomposition and effective dimension, we now address these questions, and provide simple illustrative examples in the results section, Sect. 2.6.

We know that LHS is able to maintain very good uniformity in all one dimensional projections because of its per-dimension stratification scheme. As a result of this, the variance error in integrating the one dimensional ANOVA terms $\{f_\mathbf{u} : |\mathbf{u}| = 1\}$ is very small. Using the orthogonality of ANOVA decomposition, we can write the overall function variance as

$$\sigma^2 = \sigma_1^2 + \sigma_{>1}^2: \quad \sigma_1^2 = \sum_{|\mathbf{u}|=1} \sigma_\mathbf{u}^2, \quad \sigma_{>1}^2 = \sum_{|\mathbf{u}|>1} \sigma_\mathbf{u}^2, \qquad (2.96)$$

where σ_1^2 is the variance of the one dimensional part of f given by $g_1(\mathbf{x}) = \sum_{|\mathbf{u}|=1} f_\mathbf{u}(\mathbf{x})$ and $\sigma_{>1}^2$ is the remaining variance of $f - g_1$. Also write the asymptotic variance of the standard Monte Carlo estimate as

$$\sigma_{\mathrm{MC}}^2 = \frac{\sigma^2}{n} = \frac{\sigma_1^2 + \sigma_{>1}^2}{n}. \qquad (2.97)$$

Using ANOVA decomposition, Stein [Ste87] showed that the asymptotic variance of an LHS estimate is

$$\sigma_{\mathrm{LHS}}^2 = \frac{\sigma_{>1}^2}{n} + o\left(\frac{1}{n}\right). \qquad (2.98)$$

Hence, compared to the Monte Carlo estimate (2.97), LHS achieves a variance reduction by reducing the variance in estimating the integral of the one dimensional part of f to $o(n^{-1})$.

Note that every one dimensional projection of the LHS sample is a scrambled $(0,1,1)$-net in base n. As a result, if the one dimensional part is smooth (the derivatives of $f_\mathbf{u}$ for $|\mathbf{u}|=1$ are continuous) then the second term on the right hand side of (2.98) reduces as $O(n^{-3})$, as per the results of Owen [Owe97b] regarding scrambled nets (see (2.104)). It is clear from (2.98) that this reduction is effective only if f has significant variance contribution from its one dimensional component g_1. If f has an effective dimension of 1 in the superposition sense ($s_T = 1$) then LHS is an excellent quadrature technique. Even if $s_T > 1$, many integrands have large variance contribution from their one dimensional components, explaining the success of LHS as a variance reduction technique. This result also explains why LHS is *unsuccessful* as a variance reduction technique in many settings: the integrand in those cases is probably not primarily one dimensional, because of which the gains over standard Monte Carlo are minimal.

Based on (2.98), for reasonably large sample size n, we can assume

$$\sigma_\text{LHS}^2 \approx \frac{\sigma_{>1}^2}{n} \quad \text{and} \quad \sigma_\text{MC}^2 \approx \frac{\sigma^2}{n}. \tag{2.99}$$

Then, using (2.97) we get

$$\frac{\sigma_1^2}{\sigma^2} = 1 - \frac{\sigma_{>1}^2}{\sigma^2} \approx 1 - \frac{\sigma_\text{LHS}^2}{\sigma_\text{MC}^2}. \tag{2.100}$$

We can estimate σ_LHS and σ_MC by taking the sample variance across the estimates from several LHS and Monte Carlo runs, respectively. This gives a way of using LHS to estimate the contribution of one dimensional ANOVA terms to the variance of f. We use this estimate in Sect. 2.6 to study the efficiency of LHS for different examples, and illustrate the conditions when Sobol' points perform better than LHS. For now we discuss these conditions theoretically.

The numerical results in [WS07] indicate why QMC, and Sobol' points in specific, can outperform LHS. We enumerate three types of functions for which Sobol' points can provide quadrature with improved integration errors, along with the corresponding features of Sobol' points that enable the improvement.

1) *High contribution from one dimensional ANOVA components*: This is the class of functions for which LHS provides large improvements over standard Monte Carlo. Numerical results in [WS07] show that

the discrepancies of one dimensional projections of Sobol' points are even better than for LHS. This allows us to retain the advantages that LHS provides: low variation in integration of the one dimensional parts of the integrand f.

2) *High contribution from one dimensional ANOVA components, but truncation dimension $s_T > 1$*: LHS is not able exploit small truncation dimension if it is greater than 1, since it only targets one dimensional components of the integrand. If we take any set $\mathbf{u} = \{1,\ldots,l\}$, $1 < l \ll s$ of the early dimensions of an LHS sample, the corresponding discrepancy can be as high as that for pseudorandom point sets. However, for l around 10 or less, the discrepancy of the early dimensions of a Sobol' point set can be much lower for practical sample sizes. Hence, for integrands with truncation dimension $s_T \leq 10$, Sobol' points may provide significant improvement in quadrature error, compared to LHS and random sampling.

3) *High contribution from higher dimensional ANOVA components with small truncation dimension $s_T > 1$*: This condition on f further expands the class of functions from item 2, since now we allow higher dimensional ANOVA components to have a large contribution to the function variance, as long as the corresponding dimensions are from the early dimensions of the point set. Clearly, LHS provides no extra advantage beyond that for the one dimensional projections. Sobol' points, however, do. The discrepancy of the projection of Sobol' points onto some subset $\mathbf{u}$, with small $|\mathbf{u}| > 1$, tends to be lower than that for LHS, as long as the subset is from the early dimensions; i.e., $\mathbf{u} \subset \{1,\ldots,l\}$ for small $l > 1$. These conditions on f are significantly less restrictive in practice than those for LHS quadrature being the best option, and suggest that Sobol' sequences – and any other competitive LDS – will perform better than, or as well as, LHS in general.

The reader is referred to [WS07] for some convenient mathematical constructs for the discrepancy of projections of any point set, and similar discussions using these constructs. We stress here that all the theoretical results presented here to illustrate the implications of low effective dimension for QMC are suggestive since they rely on bounds and asymptotes (e.g., the Koksma–Hlawka bound) and not exact relations. There may be cases where QMC performs well even with high effective dimension, as shown in [Tez05] for a class of functions that have full effective dimension in both the truncation and superposition senses. Tezuka shows that for these functions the QMC error decreases as $O(n^{-1})$, without the troublesome $\log^s(n)$ in the numerator, and that the

Koksma–Hlawka bound is so loose for this case as to be completely useless. This shows that low effective dimension is not necessary for QMC to beat Monte Carlo. Owen [Owe03b] points out that is also not a sufficient condition. Given these caveats however, we and several researchers [MC96][CMO97][ABG98][WF03][Owe03b] believe that low effective dimensions play a significant role in creating the conditions for improved quadrature using QMC as compared to Monte Carlo. We provide some simple examples to illustrate and support these arguments in Sect. 2.6. For now, we believe these suggestive theoretical arguments and the cited references, and propose a flow for applying QMC to statistical analysis of circuits.

2.5 Quasi-Monte Carlo for Circuits

The foregoing sections provide us sufficient information to propose a flow for applying QMC to statistical analysis of circuits. As suggested by discussions in Sect. 2.3.2, we use the Sobol' sequence as our representative LDS in the proposed flow. Once the construction of the promising Niederreiter–Xing sequences [NX96] becomes feasible, we can use them instead of the Sobol' sequence in the reasonable hope of even better performance.

2.5.1 The Proposed Flow

From Sect. 2.4.1 we know the importance of using transforms like principal components analysis to maximize the amount of variance in the inputs to the minimum number of early dimensions. Also, PCA is popularly used by researchers and practitioners in EDA [Ism93][CS05] to reduce the number of statistical parameters into a small uncorrelated set while still accounting for most of the variance of the original parameters. Hence, we assume that our QMC flow starts with post-PCA statistical parameters: this enables us to focus on aspects that are truly novel in the context of circuit analysis. In fact, if we have transformed the input sampling space to be the unit cube, then we have effectively used some orthogonal transformation like PCA to obtain independent inputs with the same variance. Another way of exploiting low effective dimension in this setting is to measure the contribution of each input of f to the variation in f, and sort the inputs in decreasing order of this measure. Such a rearrangement of the inputs helps minimize the effective dimension s_T in the truncation sense and exploit the good uniformity of the early dimensions of Sobol' points, as discussed in Sect. 2.4.2. We refer to such a rearrangement as a *variable-dimension mapping*. The impact of any input on the function can be estimated with some measure

Algorithm 2.2 QMC for statistical simulation of circuits

Require: circuit performance functions $\mathbf{f} = \{f_1, \ldots, f_{sy}\}$, joint probability distribution of inputs $\Pi(\mathbf{x})$, input dimensionality s, and sample size n

1: $\pi \leftarrow$ InputOrdering(s, $\mathbf{f}$, Π) – $\pi(j) \in \{1, \ldots, s\}$ is the j-th most important input index
2: skip $2^{\lfloor \log_2 n \rfloor}$ points of the s dimensional Sobol' sequence
3: **for** $i = 1$ to n **do**
4:    $\mathbf{z} \leftarrow$ NextSobolPoint()
5:    $x_{\pi(j)} = z_j$, $j = \{1, \ldots, s\}$
6:    $\mathbf{x_i} = \{x_1, \ldots, x_s\}$
7:    evaluate $\mathbf{y}_i = \mathbf{f}(\Pi^{-1}(\mathbf{x_i}))$
8: **end for**
9: **return** QMC sample points $\{(\mathbf{x}_i, \mathbf{y}_i)\}_{i=1}^n$

---

of global sensitivity, as in [SK05] or in Sect. 1.6.3.1 of this thesis. Here we use one of two much simpler options:

1) The designer can select the parameters that most affect the relevant performance metrics, and these can be assigned to the initial dimensions of the QMC. This can be a feasible option in manual design settings where the statistical parameters correspond to different devices in the circuit being designed, since circuit designers often have good insight regarding the devices that significantly affect the relevant performance metrics.

2) Run a small standard Monte Carlo run and compute Spearman's rank correlation coefficient between each input x_i and the circuit performance metric. Use this rank correlation as the measure of global sensitivity and sort the inputs in decreasing order of correlation before running QMC. For multiple performance metrics, use the sum of the rank correlation coefficients across all metrics with each input. Spearman's rank correlation is more robust than Pearson's linear correlation in the presence of nonlinear relationships. For an explanation of Spearman's rank correlation, please refer to Sect. 1.6.4.1. Of course, better sampling techniques, like Latin hypercube sampling, or more accurate estimates of global sensitivity, if available, may be used here. However, this simple approach also proves to be sufficiently useful, as demonstrated by the experimental results in Sect. 2.6.

Our proposed QMC algorithm is shown as Algorithm 2.2. The function InputOrdering(), shown as Algorithm 2.3, performs the global sensitivity computation to determine a permutation π such that $\pi(j)$ gives

Algorithm 2.3 The function InputOrdering() used in Algorithm 2.2

Require: circuit performance functions $\mathbf{f} = \{f_1, \ldots, f_{s_y}\}$, joint probability distribution of inputs $\Pi(\mathbf{x})$, input dimensionality s
1: $\rho_i = 0$, $i = 1, \ldots, s$
2: **for** $i = 1$ to n **do**
3:     randomly generate $\mathbf{x}_i = (x_{i1}, \ldots, x_{is})$ from Π
4:     evaluate $\mathbf{y}_i = (y_{i1}, \ldots, y_{is_y}\}$ using $y_{ij} = f_j(\mathbf{x}_i)$ for $j = \{1, \ldots, s_y\}$
5: **end for**
6: **for** $j = 1$ to s_y **do**
7:     **for** $k = 1$ to s **do**
8:         $\rho_k = \rho_k + |\text{RankCorr}(\{x_{ik}\}_{i=1}^n, \{y_{ij}\}_{i=1}^n)|$
9:     **end for**
10: **end for**
11: **return** $\pi : \{1, \ldots, s\} \rightarrow \{1, \ldots, s\}$ such that $\rho_{\pi(j)}$ is the j-th largest element in $\{\rho_k\}_{k=1}^s$

the index of the input with the j-th largest measure of global sensitivity. In our implementation, this computation involves a $n_\rho = 1{,}000$-point Monte Carlo run followed by computation of the rank correlation coefficients. Note that, in Algorithm 2.2, we skip the first $2^{\lfloor \log_2 n \rfloor}$ points of the Sobol' sequence, as recommended empirically in [ABG98] for better performance. The function NextSobolPoint() uses the smallest degree primitive polynomials and direction numbers satisfying Sobol's Property A, as discussed in Sect. 2.3.3. The function RankCorr() in Algorithm 2.3 computes Spearman's rank correlation, as per Sect. 1.6.4.1.

The sample points returned by QMC can be used for computing some metric, like circuit yield (Sect. 2.2.1.2) or the 99-th percentile, or for further analysis, like visualization or response surface modeling.

2.5.2 Estimating Integration Error

In practice, the exact value of $Q = \int f(\mathbf{x}) d\mathbf{x}$ is unknown, for example the exact value of circuit yield. Usually, this is the reason for using numerical quadrature methods. Then how do we estimate the error in the quadrature estimate Q_n? Random methods, namely Monte Carlo, make this easy since the variance of the Monte Carlo estimate can be used as a probabilistic measure of the error.

2.5.2.1 Estimating Monte Carlo Error

Theorem 2.3 in Sect. 2.2.2 shows us how, using the central limit theorem, we can approximate the distribution of the Monte Carlo error as being normal, and derive such a probabilistic measure of error. In practice, relying on this assumption of normality, we can use the sample standard deviation of the estimates from several different n-point Monte Carlo runs, to compute this probabilistic measure. Suppose we computed n_{MC} estimates $\{Q_n^{(i)}\}_{i=1}^{n_{\text{MC}}}$. The sample standard deviation is then given by

$$\hat{\sigma}_{\text{MC}}^2 = \frac{\sum_{i=1}^{n_{\text{MC}}}(Q_n^{(i)} - \bar{Q}_n)^2}{n_{\text{MC}} - 1}, \qquad (2.101)$$

where the sample mean $\bar{Q}_n$ is given by

$$\bar{Q}_n = \frac{\sum_{i=1}^{n_{\text{MC}}} Q_n^{(i)}}{n}. \qquad (2.102)$$

Then the *magnitude* of the Monte Carlo error is within

$$\hat{\sigma}_{\text{MC}} \Phi^{-1}\left(\frac{1+p}{2}\right) \qquad (2.103)$$

with probability p, where Φ is the standard normal cumulative distribution function. This corresponds to the *confidence interval* with a confidence level of p.

2.5.2.2 Estimating QMC Error with Scrambled Sequences

Quasi-Monte Carlo is a deterministic quadrature technique: we get the same estimate Q_n every time we run QMC with the same number of points, assuming no changes in the parameters of the LDS (e.g., primitive polynomials for Sobol' points). Hence, there is no natural variance that we can exploit to estimate the error as in the case of Monte Carlo. Also, bounds on the error, like the Koksma–Hlawka bound (2.26), do not help because of at least two reasons:

- It is usually computationally infeasible to estimate both $V(f)$ and D_n^* with acceptable accuracy. It should be noted here that some versions of discrepancy can be computed in reasonable time. Warnock [War72] derived an explicit formula for the L_2 star discrepancy, that was generalized in [CMO97]. See [Hic98] for some generalized error bounds. However, computing the variation of the function still remains infeasible.

- Even if the error bound can be computed, it can be very different from the actual error value, as discussed in Sects. 2.3.1 and 2.3.2.3.

One way to get around this problem is to artificially randomize the QMC points. Then, we can run randomized QMC several times and estimate probabilistic error values, just as we did for Monte Carlo. Several schemes for randomizing deterministic LDSs have been proposed, and are surveyed in [LL02]. Owen [Owe95] proposed a randomization scheme that *scrambles* (t,s)-sequences and (t,m,s)-nets while maintaining two important properties:

1) Every point in the scrambled set has a uniform distribution over C^s, so that the approximation Q_n is unbiased.

2) The resulting nets or sequences, are still (t,m,s)-nets and (t,s)-sequences in base, respectively, b with probability one, and with no change to t, m or b.

As mentioned in Sect. 2.3.4, a Latin hypercube sample falls under this class of scrambled (t,m,s)-nets. Since scrambled sequences have the two properties mentioned above, they obey the asymptotic properties and error bounds of their deterministic counterpart. Hence, we can use multiple runs with different scramblings to estimate the variance σ_{QMC} of randomized QMC and the corresponding probabilistic error estimates using (2.103), with σ_{MC} replaced by σ_{QMC}.

For theoretical results on the variance of randomized (t,m,s)-nets, see [Owe97a][Owe97b][Owe98b] by Owen. Owen shows that under certain smoothness conditions on the integrand, scrambled (t,m,s)-nets can actually achieve variance of

$$O\left(\frac{\log^{s-1}(n)}{n^3}\right), \qquad (2.104)$$

implying an asymptotic integration error rate of $n^{-1.5}$, which is even better than the standard QMC asymptotic error rate. The smoothness condition requires that the mixed partial derivative,

$$h(\mathbf{x}) = \frac{\partial^s f}{\partial x_1 \ldots \partial x_s}, \qquad (2.105)$$

satisfies the Lipschitz condition,

$$|h(\mathbf{x}) - h(\mathbf{x}')| \leq B\|\mathbf{x} - \mathbf{x}'\|_2^\beta, \qquad (2.106)$$

for some finite $B \geq 0$ and $\beta \in (0,1]$. Although we will see an example of this rate in the results section (Sect. 2.6), these properties of scrambled nets and their implications for statistical circuit analysis, are not studied in detail in this thesis and can be a fruitful target for future research.

2.5.3 Scrambled Digital (t, m, s)-Nets and (t, s)-Sequences

2.5.3.1 Owen's Scrambling

Let $\{\mathbf{x}_1, \mathbf{x}_2, \ldots\}$ and $\{\mathbf{z}_1, \mathbf{z}_2, \ldots\}$ denote the original sequence and a randomly scrambled version, respectively, both in base b. Let $x_n^{(i)}$ be the i-th coordinate of $\mathbf{x}_n$, and let its b-ary expansion be

$$x_n^{(i)} = \sum_{j=1}^{\infty} x_{i,j} b^{-j} = 0.x_{i,1} x_{i,2} \ldots, \quad i = \{1, \ldots, s\}, \tag{2.107}$$

where $x_{i,j} \in \{0, \ldots, b-1\}$ is a digit in base b. Note that we have dropped the subscript n in the expansion, to reduce notation clutter. Assume similar meanings for $z_n^{(i)}$ and $z_{i,j}$. Then,

$$z_{i,1} = \pi^i(x_{i,1}), \tag{2.108}$$

where $\pi^i : \{0, \ldots, b-1\} \to \{0, \ldots, b-1\}$ is a randomly chosen permutation for dimension i; a different such permutation is chosen randomly for each dimension. This operation is, thus, scrambling the first digit of every coordinate. Similarly, we scrambling all other digits with independent, randomly chosen permutation schemes. Furthermore, the permutation of the j-th digit depends on the precise values of the previous $j-1$ digits. We write this as

$$z_{i,j} = \pi^i_{x_{i,1}, x_{i,2}, \ldots, x_{i,j-1}}(x_{i,j}). \tag{2.109}$$

For example, the permutation scheme of the third bit in 0.111 will be separate from the permutation scheme of the third bit in 0.101 even though the value of the third bit is the same in both cases. This is because the entire sequence of bits before the third bit determines the permutation applied to the third bit.

Such a scrambling scheme can be computationally tedious because of the extensive bookkeeping required: the number of permutations is $s \frac{b^m - 1}{b - 1}$ for m digits. Less expensive scrambling schemes have been proposed in [Mat98][FT02][Owe03a], among others. Owen [Owe03a] also studies the variance of quadrature estimates from these different scrambling schemes. We use the *linear matrix scrambling* method for digital nets and sequences, as implemented in [HH03].

2.5.3.2 Linear Matrix Scrambling: A Simpler Scheme

Let us rewrite the digital construction of (2.52) in current notation. For integer $n = 1, 2, \ldots$, the b-ary expansion of $n-1$ is

$$n - 1 = \sum_{k=0}^{\infty} a_k b^k = \ldots a_2 a_1 a_0. \tag{2.110}$$

Then the digital construction of the i-th coordinate of n-th point is

$$\begin{pmatrix} x_{i,1} \\ x_{i,2} \\ \vdots \end{pmatrix} = \mathbf{C}^{(i)} \begin{pmatrix} a_0 \\ a_1 \\ \vdots \end{pmatrix} \pmod{b}, \tag{2.111}$$

where we have suppressed n to reduce clutter. The linear matrix scrambling method modifies this construction as

$$\begin{pmatrix} z_{i,1} \\ z_{i,2} \\ \vdots \end{pmatrix} = \mathbf{L}^{(i)} + \mathbf{e}^{(i)} \begin{pmatrix} x_{i,1} \\ x_{i,2} \\ \vdots \end{pmatrix}$$

$$= \mathbf{L}^{(i)} \mathbf{C}^{(i)} \begin{pmatrix} a_0 \\ a_1 \\ \vdots \end{pmatrix} + \mathbf{e}^{(i)} \pmod{b}, \quad i = 1, \ldots, s. \tag{2.112}$$

Here, $\mathbf{L}^{(i)}$ are randomly and independently chosen non-singular lower-triangular matrices over $Z_b = \{0, \ldots, b-1\}$, of size $\infty \times \infty$ for infinite precision, and $m \times m$ for m digits of precision. $\mathbf{e}^{(i)}$ are randomly and independently chosen vectors over Z_b, of length same as $\mathbf{L}^{(i)}$. The n-th scrambled point is

$$\mathbf{z}_n = (0.z_{1,1}z_{1,2}\ldots, 0.z_{2,1}z_{2,2}\ldots, \ldots, 0.z_{s,1}z_{s,2}\ldots). \tag{2.113}$$

This scrambling method is clearly much simpler than Owen's scrambling, but is also less rich in its range of random permutations. For example, from (2.112), the first digit $z_{i,1} = l_{11}^{(i)} x_{i,1} + e_1^{(i)}$, where $l_{11}^{(i)}$ and $e_i^{(i)}$ are the first elements of $\mathbf{L}^{(i)}$ and $\mathbf{e}^{(i)}$, respectively. $l_{11}^{(i)} \in \{1, \ldots, b-1\}$ for non-singular $\mathbf{L}^{(i)}$, and $e_1^{(i)} \in \{0, \ldots, b-1\}$, giving us $b(b-1)$ possible permutations, while in Owen's method we have $b!$ possibilities. However, this smaller range of possibilities is not too restrictive and suffices for our experiments. We now show how this scrambling technique can be incorporated into Sobol's construction using direction numbers.

2.5.3.3 Scrambling Sobol' Sequences with Linear Matrix Scrambling

We know from (2.59) that the j-th column of $\mathbf{C}^{(i)}$ contains the bits of j-th direction number $v_j^{(i)}$ for dimension i. Let $\mathbf{v}_j^{(i)}$ denote the vector of the bits of $v_j^{(i)}$; i.e., $\mathbf{v}_j^{(i)}$ is the j-th column of $\mathbf{C}^{(i)}$. If we write

$$\mathbf{L}^{(i)}\mathbf{C}^{(i)} = \mathbf{L}^{(i)}[\mathbf{v}_1^{(i)}\ \mathbf{v}_2^{(i)}\ \ldots] = [\mathbf{v'}_1^{(i)}\ \mathbf{v'}_2^{(i)}\ \ldots], \qquad (2.114)$$

then, denoting $l_{jk}^{(i)}$ as the (j,k)-th element of $\mathbf{L}^{(i)}$, and $v_{j,k}^{(i)}$ as the k-th bit in $\mathbf{v}_j^{(i)}$, we get

$$\mathbf{v'}_j^{(i)} = \begin{pmatrix} l_{11}^{(i)}v_{j,1}^{(i)} + l_{12}^{(i)}v_{j,2}^{(i)}\ldots \\ l_{21}^{(i)}v_{j,1}^{(i)} + l_{22}^{(i)}v_{j,2}^{(i)}\ldots \\ \vdots \end{pmatrix}. \qquad (2.115)$$

Let $\mathbf{l}_j^{(i)} = (l_{1j}^{(i)}, l_{2j}^{(i)}, \ldots)^T$ be the j-th column of $\mathbf{L}^{(i)}$. Then, using bitwise Boolean operations we can write

$$\mathbf{v'}_j^{(i)} = v_{j,1}^{(i)} \cdot \mathbf{l}_1^{(i)} \oplus v_{j,2}^{(i)} \cdot \mathbf{l}_2^{(i)} \oplus \cdots, \quad 1 \leq i \leq s,\ j > 0. \qquad (2.116)$$

This bit vector, $\mathbf{v'}_j^{(i)}$, is the j-th column of the i-th *scrambled* generator matrix given by $\mathbf{L}^{(i)}\mathbf{C}^{(i)}$, and corresponds to a new scrambled direction vector $v'^{(i)}_j$ for the Sobol' construction. Then, corresponding to (2.60), the i-th coordinate for the n-th point is given by

$$z_n^{(i)} = a_0 v'^{(i)}_1 \oplus a_1 v'^{(i)}_2 \oplus \cdots \oplus a_{m-1} v'^{(i)}_m \oplus \mathbf{e}^{(i)}. \qquad (2.117)$$

If we use the Gray code construction as in (2.79), we need to XOR $\mathbf{e}^{(i)}$ only once, to the first point, and subsequent points are given simply as

$$z_{n+1}^{(i)} = z_n^{(i)} \oplus v'^{(i)}_l, \qquad (2.118)$$

where l is the index of the bit where the Gray codes of $n-1$ and n differ. We are now well-equipped to demonstrate the performance of QMC using experiments. We do this in the next section.

2.6 Experimental Results

In Sect. 2.4.2 we concluded, based on theoretical considerations, that QMC using Sobol' points should result in smaller errors and possibly faster convergence, when compared to Latin hypercube sampling. We now test this conclusion experimentally on some simple examples that also allow us to validate the results analytically. After this, we demonstrate the performance of QMC on a variety of circuit benchmarks, in comparison with standard Monte Carlo and LHS.

2.6.1 Comparing LHS and QMC (Sobol' Points)

Let f be our integrand. We test two conclusions from Sect. 2.4.2 here:

1) LHS almost completely and exclusively removes the variance contribution of the one dimensional ANOVA components of f, to achieve variance reduction. Hence, we can estimate the variance contribution of the one dimensional components via (2.100). See Sect. 2.4.1 for a discussion on the ANOVA decomposition.

2) LHS restricts its variance reduction activity to the one dimensional components of f. Sobol' points provide further benefit by reducing the error in integrating also some higher dimensional components, because they enjoy highly uniform higher dimensional projections in the early dimensions.

2.6.1.1 LHS (Almost) Exactly Removes One Dimensional Variance Contribution

Consider the following three functions in five dimensions.

1) *An additive function* with only one dimensional nonzero components; i.e., with effective dimension s_S equal to 1, in the superposition sense.

$$f_a = x_1^2 + x_2^2 + x_3^2 + x_4^2 + x_5^2. \tag{2.119}$$

Note that the function has full truncation dimension $s_T = 5$. We expect LHS to almost completely remove any variance in the integral estimate for this function, because there is no contribution from any multi-dimensional components.

2) *A cross-term function* with significant contributions from multi-(two-)dimensional components; i.e., with superposition dimension $s_S = 2$. The truncation dimension is still 5.

$$f_c = (x_1 + x_2 + x_3 + x_4 + x_5)^2. \tag{2.120}$$

Note that f_a is part of f_c. We expect the effectiveness of LHS to be less for this function, and that the remaining variance in the estimate (σ_{LHS}^2) is proportional to the variance contribution from the two dimensional components.

	f_a	f_c	f_s
$\hat{\sigma}^2_{\text{MC}}$	4.024×10^{-5}	9.483×10^{-4}	6.112×10^{-4}
$\hat{\sigma}^2_{\text{LHS}}$	5.284×10^{-13}	2.839×10^{-5}	2.839×10^{-5}
$\hat{\eta}_{\text{LHS}} = \widehat{\frac{\sigma_1^2}{\sigma^2}} = 1 - \frac{\hat{\sigma}^2_{\text{LHS}}}{\hat{\sigma}^2_{\text{MC}}}$	1.000	0.970	0.954
Exact σ^2	–	$193/18$	$125/18$
Exact σ_1^2	–	$94/9$	$20/3$
$\eta = \frac{\sigma_1^2}{\sigma^2}$	1	0.974	0.960

Table 2.3. Fractional variance contribution from one dimensional components of f_a, f_c and f_s computed using LHS estimate (2.100) and analytically. We see that LHS does exclusively remove the variance from one dimensional components

3) *A strongly cross-term function* with even higher relative contribution from two dimensional components.

$$f_s = f_c - f_a = 2 \sum_{i=1}^{4} \sum_{j=i+1}^{5} x_i x_j. \qquad (2.121)$$

We expect that the variance of LHS, σ^2_{LHS} will not change from the case of f_c since we have only removed the one dimensional components and not changed anything in the two dimensional components.

We ran 30 Monte Carlo runs and 30 LHS runs, each with a sample size of $n = 10{,}000$, to estimate the Monte Carlo variance (σ^2_{MC}) and the LHS variance (σ^2_{LHS}), respectively. We use the sample variance formula (2.101). Plugging these variance estimates into (2.100), we can estimate the fraction of variance contributed by the one dimensional components of f. Let σ^2 be the total function variance, σ_1^2 the variance from the one dimensional components and $\eta = \sigma_1^2/\sigma^2$ the fraction of variance from one dimensional components. Table 2.3 shows the results in data rows 1–3. We can see that, as expected, the LHS variance for the additive function f_a is negligibly small. Since f_a has only one dimensional components, all its variance is from one dimensional components. The estimate $\hat{\eta}_{\text{LHS}}$ is almost exact ($=1$). For the other functions, we can analytically compute σ^2, σ_1^2 and η. These exact values are shown in data rows 4–6. Derivations are given in Appendix A. We see that the estimates $\hat{\eta}_{\text{LHS}}$ are very close to the exact values, providing strong evidence for the claim that LHS exclusively removes σ_1^2 from the estimate variance.

Quasi-Monte Carlo

2.6.1.2 Sobol' Points Are Better Than LHS for Functions with Significant Higher Dimensional Components

All three test functions above allow a simple analytical computation of their exact integrals over the unit cube C^s, this being one of the reasons for choosing them as test functions. These exact values are $Q(f_a) = 5/3$, $Q(f_c) = 20/3$ and $Q(f_s) = 5$. We also computed these integrals numerically in three different ways:

1) Using standard Monte Carlo with increasing number of points n. The values of n are chosen to match those for the LHS samples sizes below.

2) Using Latin hypercube samples with sample sizes of $n = 100 \cdot 2^{\{0,\ldots,7\}} = \{100, 200, 400, \ldots, 12{,}800\}$.

3) Using Sobol' point with the same samples sizes as LHS.

Since we know the exact answers, we can directly compute the relative error without having to resort to probabilistic errors based on sample variance. The plots in Fig. 2.8 show the relative integration error for these three methods on all three test functions on a $\log_{10} - \log_{10}$ scale. Least squared-error linear fits in this scale, shown as dashed straight lines, estimate the convergence exponent of each integration method as the slope of the fit. These estimated rates are annotated on the corresponding linear fits. We now discuss each of the three test cases is some detail, in the context of these results.

- *Additive function f_a (Fig. 2.8(a))*: Monte Carlo achieves an estimated error rate of $n^{-0.4963}$, which is close to the expected $n^{-0.5}$ rate. The Sobol' points achieve a rate of $n^{-0.8963}$, which is close to the asymptotic rate of n^{-1} for QMC. This suggests that the truncation and superposition dimensions ($s_T = 5, s_S = 1$) are small enough for the Sobol' points to exploit. Interestingly, LHS achieves a *much* faster convergence of $n^{-1.5702}$ along with lower error. This superiority of LHS over Sobol' points is actually expected. We know, from Sect. 2.3.4, that an LHS sample is a scrambled (t, m, s)-net. Since the function f_a satisfies Owen's smoothness condition (2.106), the result in (2.104) predicts an asymptotic $n^{-1.5}$ convergence for LHS error. Similar behavior is predicted by Fox in [Fox99], Theorem 9.1.2.

- *Cross-term function f_c (Fig. 2.8(b))*: Even with contribution from higher dimensional components in f, Monte Carlo achieves more or less the same convergence rate ($n^{-0.4383}$), close to the expected, asymptotic $n^{-0.5}$. However, we see a big change in the performance

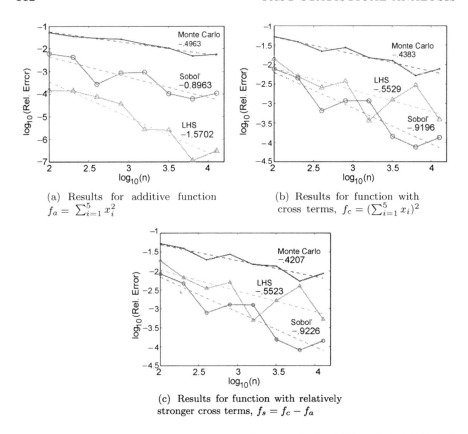

Figure 2.8. Comparison of relative errors of Monte Carlo, LHS and QMC (Sobol' points) with increasing number of points. The three test functions have different relative contributions from their one dimensional ANOVA components

of LHS. It provides no benefit over Monte Carlo in integrating the *two* dimensional components of f, resulting in an overall error rate of $n^{-0.5529}$, which is closer to the Monte Carlo error rate. Note that it is still significantly better than Monte Carlo because its excellent performance on the one dimensional components. Interestingly, the Sobol' points maintain their low error and fast convergence in spite of the increased superposition dimension of the integrand. This supports our argument that Sobol' points have good uniformity in higher dimensional projections of the early dimensions, which allows them to integrate the higher dimensional ANOVA components with better accuracy than Monte Carlo or LHS.

- *Strong cross-term function f_s (Fig. 2.8(c))*: This function has even higher relative contribution from its two dimensional components. However, we see no significant difference in the performance of any of the three methods. This is not surprising. The variance of Monte Carlo, of course, does not exploit any ANOVA features of the integrand. This explains the lack of change in its performance. LHS primarily targets the one dimensional components and has Monte Carlo-type performance on higher dimensional components of f_s. f_c and f_s differ only in their one dimensional components and have identical two dimensional components. Hence, the error in the LHS estimate, which is almost completely due to the latter, does not change. Similar arguments apply to the Sobol' points: the change in the error due to changes in the one dimensional components is very small.

Based on these experiments and the arguments in Sect. 2.4.2, we can confidently conclude that, in the general case, Sobol' sequences will show lower error and faster convergence than LHS or Monte Carlo, as long as the truncation dimension of the integrand is not too large. The only exception is when almost all of the variance contribution is due to the one dimensional components of the integrand and the integrand is smooth, in which case LHS or scrambled QMC will perform better. For the case of statistical circuit analysis, we neither expect the integrand to be primarily one dimensional, nor to be smooth (e.g., the characteristic function integrand (2.12) for circuit yield). However, from common design knowledge, we believe that the truncation dimension of the integrand will not be too large. Hence, Sobol' points (or any other competitive QMC method) seem an appropriate choice.

2.6.2 Experiments on Circuit Benchmarks

We now demonstrate the performance of QMC on circuit benchmarks. Before we discuss the benchmarks and the results, we briefly mention some relevant implementation details. A linear congruential generator (LCG) [Gla04] (`drand48()` in C) was used to generate the pseudorandom sequences for standard Monte Carlo and for random scramblings of the Sobol' points. This generator enjoys widespread popularity and the obtained results will be immediately relevant to the general practitioner. Also, variance results in [OE04] comparing LCG with a generalized feedback shift register generator (GFSR) [MK94], do not show significant improvements for GFSR in the context of randomized QMC. The standard Box Muller [BM58] method for generating normally distributed variates is inaccurate, especially for a large number of samples [Tez95]. Hence, an inverse transform method, by Acklam [Ack], was used. This is the Π_i^{-1} in (2.17) for the case of normal variates. Now we describe the benchmark circuits and the experiments. All results will be discussed together after this description. We use the following three benchmarks.

1) **Master–slave flip-flop with scan chain (MSFF)**: This is the same circuit used as a benchmark for SiLVR in Sect. 1.7.1. In this case, we are computing the parametric yield, given a maximum acceptable clock-output delay (τ_{cq}) of 200 ps. This is a 31 dimensional problem. 10 Monte Carlo runs with 50,000 points each were run to compute the Monte Carlo variance. The QMC variance is computed across a set of 10 runs: one run using 50,000 Sobol' points and 9 runs using 50,000 scrambled Sobol' points each. Each scrambled run uses a distinct set of permutations.

2) **Sub-1 V CMOS bandgap voltage reference**: This benchmark is also used for testing SiLVR in Sect. 1.7.3, where a detailed description is also provided. In this case, we compute the parametric yield, given three specifications: 1) output voltage, V_{ref} within 10% of 600 mV, 2) output settling time $\tau_s \leq 200$ ns, and 3) dropout voltage $V_{do} \leq 900$ mV. The settling time is defined as the time taken by the output to settle within 1% of its final value. This is a 122 dimensional problem. We use the same run plan as for the flip-flop benchmark, but with a sample size of 20,000 for each run.

3) **64-bit SRAM column**: This benchmark is also used for testing Statistical Blockade in Sect. 3.3.4, where a detailed description is also given. In this case, we are computing the 90-th percentile of the write time τ_w in the presence of manufacturing variations. This is a 403 dimensional problem. The same run plan as for the flip-flop is used, with the only difference being the sample size for each run: here we use 10,000 points.

It is clear that the problem dimensions are large enough such that 10,000–50,000 Sobol' points will not be uniformly distributed over all dimensions. We are bound to get undesirable patterns in several projections, similar to the ones shown in Sect. 2.4. Here, it becomes important that we use some technique to reduce the effective dimension of the problem. As described in Sect. 2.5.1, we use Spearman's rank correlation (1.69) as a measure of variable importance (or global sensitivity), and arrange the statistical parameters in decreasing order of importance, before running QMC.

As an illustrating example, let us look at how the rank correlation based variable-dimension mapping works for the flip-flop, shown in Fig. 2.9(a). Figure 2.9(b) shows the magnitude of the rank correlation ($|\rho_S|$) of each parameter with the clock-output delay for rising output, computed from an initial Monte Carlo run of 1,000 samples. The variables are sorted in *decreasing* order of importance (rank correlation mag-

Quasi-Monte Carlo

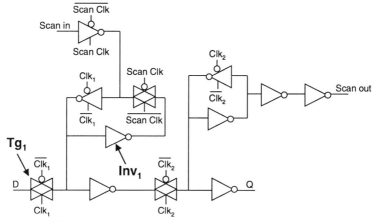

(a) Master–slave flip-flop with scan chain component

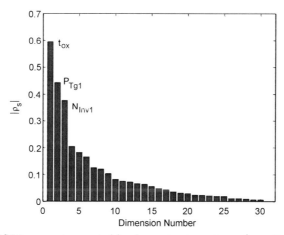

(b) MSFF parameters sorted by decreasing importance (magnitude of rank correlation with the flip-flop delay)

Figure 2.9. These figures illustrate the use of rank correlation as a measure of parameter importance, to be used for variable-dimension mapping

nitude): this is now the order they will be mapped to the increasing dimensions of the Sobol' sequence. The three most important parameters are labeled: 1) t_{ox}: global gate oxide variation, 2) P_{Tg1}: the V_t variation in the pMOS device in the input transmission gate Tg_1, and 3) N_{Inv1}: the V_t variation in the nMOS device in the inverter Inv_1. The latter two devices are on the critical signal path for a high input causing a rising output, and are important for correctly sampling a "1" at the input, especially when the input timing is close to the setup limit. Since the input was timed in such a manner in the testbench, these measures of importance make intuitive sense.

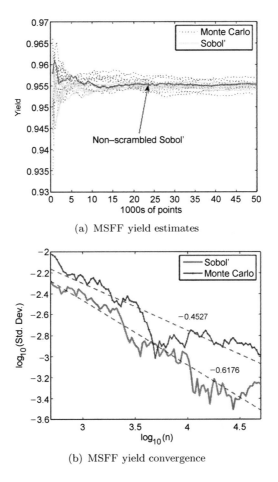

Figure 2.10. Comparison of Monte Carlo and QMC estimates and std. deviation convergence, for the flip-flop benchmark

2.6.2.1 Analysis of Results

Figures 2.10(a), 2.11(a) and 2.12(a) plot the values of the estimates with increasing number of points for each Monte Carlo (pseudorandom) and QMC (Sobol') run. For all three cases, we can clearly see that the QMC graphs converge more quickly than the Monte Carlo graphs in general. In particular, the *non-scrambled* Sobol' points converge very fast towards the final result. This fact provides indirect validation that our rank correlation based dimension mapping is an effective heuristic. Scrambling the digits of an LDS sample changes the way the sampling space is filled up, and hence, changes the patterns and the discrepancies of the projections of the point set. We observe here that changing the patterns in this way causes the QMC performance to degrade

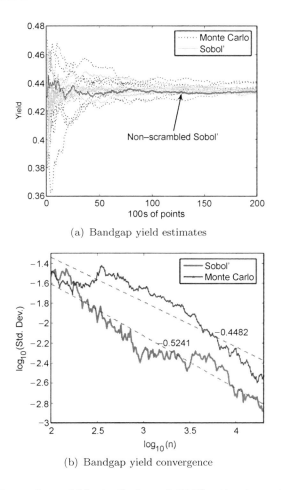

Figure 2.11. Comparison of Monte Carlo and QMC estimates and std. deviation convergence, for the voltage reference benchmark

in general for our benchmarks. This implies that the rank correlation arranges the variables in a way that is optimal (or at least, advantageous), given the patterns of the non-scrambled LDS. This behavior is more pronounced as the problem dimensionality increases from MSFF to the SRAM Column, suggesting that for low dimensionality (e.g., 31 dimensional MSFF), the LDS uniformity does not show large variation for different projections. For high dimensional problems, however, effective variable-dimension mapping should give notable improvement, over a random or uneducated assignment of variables to LDS dimensions. Of course, the impact of such mappings depends on the minimum possible truncation dimension of the integrand. If the truncation dimension of a problem cannot be made much smaller than the full di-

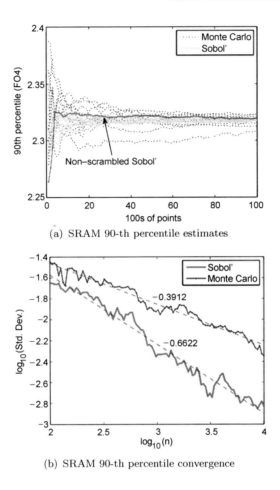

(a) SRAM 90-th percentile estimates

(b) SRAM 90-th percentile convergence

Figure 2.12. Comparison of Monte Carlo and QMC estimates and std. deviation convergence, for the SRAM column benchmark

mensionality, then all mappings will achieve similar performance. Again, in high dimensions, it is very likely that the minimum possible truncation dimension will be much smaller than the full dimensionality. As a result, using the correct mapping will result in much improved performance.

Figures 2.10(b), 2.11(b) and 2.12(b) compare the standard deviation of the Monte Carlo runs ($\hat{\sigma}_{MC}$) and the QMC runs ($\hat{\sigma}_{QMC}$) with increasing number of points, showing the effectiveness of scrambled QMC as a variance reduction method. The plots are in $\log_{10} - \log_{10}$ scale, where a $\sigma \propto n^{-\alpha}$ relationship will appear as a straight line with slope $-\alpha$, similar to the plots in Fig. 2.8. Linear fits, via least squared error, are shown as dashed straight lines, and are annotated with the corresponding convergence exponent. We can see that, in general, QMC shows lower variance and faster convergence than Monte Carlo across all three benchmarks. The estimated Monte Carlo convergence rates are a little slower than the asymptotic rate of $n^{-0.5}$. This can be because we have not reached the asymptotic rate and also because the estimates are computed from only 10 runs. A larger number of runs is definitely desirable, but eludes us because of the large circuit simulation times. Even with these approximate estimates we do get a good sense of the performance of QMC relative to Monte Carlo. σ_{QMC} shows convergence rates in between the Monte Carlo and QMC asymptotic rates of $n^{-0.5}$ and n^{-1}, respectively. This suggests that the integrands for these benchmarks have superposition dimension greater than 1 and moderately large truncation dimension. Another reason for these reduced rates can be the lack of smoothness in characteristic function integrand (2.12). Integrand smoothness can lead to better QMC performance as indicated in [MC96] and [Fox99]. Figure 2.13 provide evidence for larger than one superposition dimension. Here we plot the standard deviation of LHS estimates for the SRAM column and voltage reference cases. We can see that the LHS curve lies in between the curves for Monte Carlo and QMC. This indicates that there are some significant multi-dimensional components of the integrands for which the Sobol' points further reduce the integration error over LHS, similar to the case of functions f_c and f_s in Fig. 2.8.

We now compute some estimates of the samples size needed to achieve a given accuracy criterion. Say the exact value of the integral is Q and we specify an accuracy criterion as follows: we want the estimate Q_n to be within $\delta\%$ of Q with a probability of p. In other words, we want the error magnitude to be less than or equal to $Q(\frac{\delta}{100})$. Using the estimate for probabilistic error from (2.103), we can write this as

$$\hat{\sigma}_{\text{MC}} \Phi^{-1}\left(\frac{1+p}{2}\right) \leq Q\left(\frac{\delta}{100}\right). \tag{2.122}$$

For $p = 0.9545$, we get

$$\hat{\sigma}_{\text{MC}} \leq Q\left(\frac{\delta}{200}\right). \tag{2.123}$$

Using the linear fits from Figs. 2.10(b), 2.11(b) and 2.12(b), we can then estimate the number of points n needed to satisfy this criterion. The same arguments hold for σ_{QMC} and the QMC sample size. The results for $\delta = 1, 0.1$ are shown in Table 2.4. Since, we do not know Q for these circuit benchmarks, we estimate it using *all* the points at our disposal – from 10 Monte Carlo and 10 QMC runs – and assume that the error in this estimate is negligible in comparison with $Q(\frac{\delta}{100})$. Even

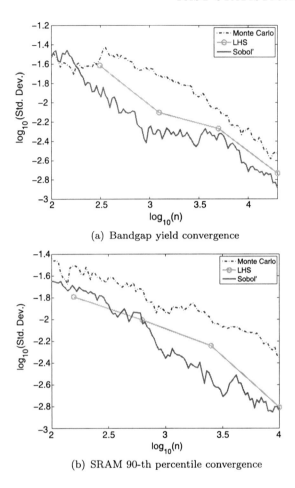

Figure 2.13. Comparison of std. deviation convergence of Monte Carlo, LHS and QMC

if the error is not negligible, because we are using the same assumption for both Monte Carlo and QMC, the *relative* trends seen here can be believed. We can see moderate to large speedups ($2\times$ to $50\times$), showing the effectiveness of scrambled QMC as a variance reduction method. Furthermore, these speedups tend to improve as the required accuracy increases.

The results presented in this section are promising and recommend using QMC for general circuit analysis problems. Of course, these are initial results and there is much scope for more research in this area. We discuss some immediate directions for future work in the following section.

Quasi-Monte Carlo

δ	MSFF	SRAM column	Voltage ref.
	MC / QMC	MC / QMC	MC / QMC
1%	1,114 / 588 (1.9×)	1,631 / 354 (4.6×)	89,115 / 10,360 (8.6×)
0.1%	180,232 / 24,465 (7.4×)	586,771 / 11,451 (51.2×)	15,182,252 / 838,062 (18.1×)

Table 2.4. Number of points needed to achieve a given error with a confidence level of 95.45%. Speedup of QMC over MC is shown in brackets

2.7 Future Work

This study brings up many relevant questions and possibilities. We outline a few here.

1) Owen proposes *Latin supercube sampling* (LSS) in [Owe98a]. LSS combines LHS and QMC in an attempt to exploit the excellent properties of QMC for the small dimensional projections, while achieving at least Monte Carlo-type performance for the high dimensional projections. The set of s dimensions is divided into k exclusive subsets. Scrambled QMC is used for each subset, with different scramblings for each subset. Then, the points in each subset are randomly permuted, as in LHS, before combining them together to achieve an s dimensional LSS sample. Each subset enjoys the scrambled QMC rate of convergence and the variance resulting from the interaction between subsets enjoys the Monte Carlo rate. We can see the similarity with LHS, where each subset is of size one. In practice, the number of statistical parameters in a circuit can become extremely large (1,000s or more). In such cases, it will likely be essential to use such mixed sampling methods to achieve effective performance from QMC. Spanier [Spa95] suggests a less powerful, but easier to apply, hybrid sampling technique using QMC for the first d dimensions and Monte Carlo for the rest.

2) We saw theoretical results in Sect. 2.5.2 and experiments (on LHS) in Sect. 2.6.1 that suggest that scrambled QMC can achieve up to $n^{-1.5}$ error convergence asymptotically, if the integrand is smooth. Morokoff and Caflisch [MC95] show that the lack of continuity in the integrand can reduce the effectiveness of non-scrambled QMC, resulting in Monte Carlo type performance. Integrands in circuit yield analysis are typically characteristic functions as in (2.12), which are discontinuous at the boundary of the acceptance region where they

suddenly change from 1 to 0. For any arbitrary boundary, whether a QMC point falls within the boundary to contribute a 1 to the integral, or outside the boundary to contribute a 0, is essentially random. This random sampling around the entire boundary leads to the degradation in QMC performance towards Monte Carlo performance. In high dimensions, the boundary becomes relatively more significant (e.g., the ratio of the boundary area of a unit cube to its volume, in s dimensions is $2s$), and the degradation worsens with increasing dimensions. Moskowitz and Caflisch [MC96] shows a method of "smoothing" such integrands by enforcing continuity, without changing the value of the integral. Fox [Fox99] discusses other forms of smoothing in the context of randomized QMC. Using such, or novel, smoothing techniques can help further improve the performance of QMC for circuits with medium dimensionality, and make QMC effective on problems with very large dimensionality.

3) Variance reduction techniques [Gla04][Fis06] are widely employed to reduce the variance – and, hence, the error – of standard Monte Carlo. Since QMC shows Monte Carlo type performance on integrands with large effective dimension, variance reduction techniques like control variates, stratification and importance sampling should be very useful in such cases. In fact, LSS is a form of stratified sampling applied to QMC. Some applications of these techniques are discussed in [Fox99].

Chapter 3
Statistical Blockade: Estimating Rare Event Statistics

3.1 Motivation

Consider the case of a 1 megabit (Mb) SRAM array, which has 1 million "identical" instances of an SRAM cell. These instances are designed to be identical, but due to manufacturing variations, they usually differ. Suppose we desire a *chip* yield of 99%; that is, no more that one chip per 100 should fail. This means that on average, not more than (approx.) one per 100 × 1 million SRAM cells; that is 10 per billion, should fail. This translates to a required *circuit* yield of 99.999999%, or a maximum *failure rate* of 0.01 ppm for the SRAM cell. This failure probability is the same as for a 5.6σ point on the standard normal distribution. If we want to estimate the yield of such an SRAM cell in the design phase, a standard Monte Carlo approach would require at least 100 million SPICE simulations on average to obtain just one failing sample point! Even then, the estimate of the yield or failure probability will be suspect because of the lack of statistical confidence, the estimate being computed using only one failing example. Such a large number of simulation is utterly intractable. This example clearly illustrates the widespread problem with designing robust memories in the presence of process variations: we need to simulate *rare* or *extreme* events and estimate the statistics of these rare events. The problem of simulating and modeling rare events stands for any circuit that has a large number of identical replications on the same chip, as in DRAM arrays and non-volatile memories. We term such circuits as *high replication circuits* (HRCs).

Note that systematic variations (e.g., proximity-based lithographic effects) can be well accounted for in SRAM cells, because they are typically small in size: the ubiquitous 6T SRAM cell contains only six transistors.

What really cause significant variation then are the random inter-device variation sources, like RDF and RCO (Sect. I.1). Further, the impact of variations is roughly inversely proportional to the square root of the transistor area [PDW89], and the transistors in SRAM cells tend be of minimum size. Hence, SRAM cells are particularly susceptible to these random variations, increasing the need for an efficient yield estimation technique for these cells.

Memory designers have typically side-stepped the problem of yield estimation by using multiple process and environmental corners with large safety margins. This approach, of course, is unreliable since it does not account for the actual statistics of the SRAM cell performance metrics. Worse, it usually results in significant over-design, which translates to a squander of chip area and power, both being expensive commodities. Monte Carlo simulation would be the ideal technique for reliably estimating the yield, but as we saw, it can be prohibitively expensive for HRCs. One avenue of attack is to abandon Monte Carlo. Several analytical and semi-analytical approaches have been suggested to model the behavior of SRAM cells [BTM01][MMR04][CC05] and digital circuits [MMR05] in the presence of process variations. All suffer from approximations necessary to make the problem tractable, or apply to a specific performance metric. [MMR04] and [MMR05] assume a linear relationship between the statistical variables and the performance metrics (e.g., static noise margin (SNM)), and assume that the statistical process parameters are normally distributed. These assumptions result in a normal distribution assumption for the performance metric too, which can suffer from gross errors, especially while modeling rare events: we shall see examples in the results section. When the distribution varies significantly from Gaussian, [MMR04] chooses an F-distribution in an ad hoc manner. [BTM01] presents a complex analytical model limited to a specific long-channel transistor model (the transregional model) and further limited to only static noise margin analysis for the 6T SRAM cell. [CC05] again models only the static noise margin (SNM) for sub-threshold SRAM cells under assumptions of independence and identical distribution of the upper and lower SNM, which may not always be valid. All these methods are specific to either one circuit, or one device model, or one performance metric. This is a general problem with analytical methods: they are not generalizable.

A different avenue of attack is to modify the Monte Carlo strategy. [HLT83] shows how importance sampling can be used to predict failure probabilities. Recently, [KJN06] applied an efficient formulation of these ideas for modeling rare failure events of single 6T SRAM cells, based on the concept of *mixture importance sampling* from [Hes03]. The approach uses real SPICE simulations with no approximating equations. However, the method only estimates the failure (exceedance) probability of a *single* threshold value of the performance metric. A re-run is needed to obtain probability estimates for another failure threshold: no complete model of the tail of the distribution is computed. The method also combines all performance metrics to compute a failure probability, given fixed thresholds. Hence, there is no way to obtain separate probability estimates for each metric, other than a separate run per metric.

In this chapter, we develop a novel, general and efficient Monte Carlo method that addresses both of the problems previously mentioned: very fast generation of 1) rare event samples, and 2) sound models of the rare event (distribution tail) statistics for any performance metric. We refer to this method as *statistical blockade* (SB). It imposes almost no a priori limitations on the form of the statistics for the statistical parameters, device models or performance metrics. The method is conceptually simple and employs ideas from two rather non-traditional sources: *extreme value theory* and *machine learning*.

Extreme value theory (EVT) [Res87] is a branch of probability that studies and optimally quantifies the statistics of, as the name suggests, extreme or rare events. It has found wide statistical application in fields such as hydrology [dH90], insurance [EKM03] and finance [EKM03] among several others: wherever there is a need to estimate the probability of rare events. One of the most consequential applications, and, indeed, one of the driving forces for the development of the theory of extremal statistics, was the Dutch dike project following the disastrous North Sea flood of 1953 that took over 1,800 human lives. One aspect of the post-flood response was to determine appropriate heights for the sea dikes in the Netherlands, such that the probability of a flood in a year is reduced to some very small amount (e.g., 10^{-4}). Technically, this involved estimating the height of sea-water level corresponding to this probability level – definitely a rare event – using statistical inference based on historical data of sea-water level measurements. Furthermore, the quantile to be estimated was much beyond the available data range. Our problem of estimating extreme quantiles of the SRAM static noise margin, using a limited number of Monte Carlo samples, is similar in flavor to the dike height problem (if not in impact on the human condition). Hence, we can employ the same technical tools from EVT for our problem.

However, the yield estimation problem is more "extreme" in the sense of the failure probabilities to be estimated: often 10^{-8} to 10^{-9} or smaller. To achieve reliable estimates of these quantiles we can need, again, impractically large Monte Carlo sample sizes. We tackle this problem by using a *filter* to intelligently simulate only those points that are important; i.e., rare. We note that *generating* each Monte Carlo point is neither challenging, nor expensive relative to *evaluating* it using a SPICE simulation. Hence, we use an appropriate filter to *block* those points that are unlikely to fall in the low-probability tails of the performance metrics. Many points are generated, but only the "rare" events are simulated. Such a partial sampling of the performance distributions fits well with the results from EVT that we exploit. The filter we use is a standard *classifier* from machine learning, and its "blocking" activity gives the method its name of statistical blockade.

In the rest of the chapter, we review relevant results from EVT, highlighting the limit theorems for the distributions of rare events. Then we show how we can use these results for statistical inference from data, which in our case is generated from a Monte Carlo simulation. Some background on classifiers is discussed, allowing us to develop the proposed statistical blockade framework. We then show how to extend this framework to metrics with conditionals (e.g., max(), min()) that result in disjoint rare event regions in the statistical parameter space, along with a recursion based extension to produce reliable estimates for extremely rare events (6 to 8σ). Finally, we present experimental results demonstrating the effectiveness of statistical blockade on realistic, circuit test cases.

3.2 Modeling Rare Event Statistics

Rare events and their statistics have been deeply studied in the fields of probability, reliability, hydrology and actuarial science. Let us first state our modeling problem concretely and compile some theoretical results that will help us solve this problem.

3.2.1 The Problem

Suppose we want to model the rare event statistics of the write time of an SRAM cell. Figure 3.1 shows an example of the distribution of the write time. We see that it is skewed to the right with a *heavy* right tail. A typical approach is to run a Monte Carlo with a small sample size (e.g., 1,000) and fit a standard analytical distribution to the data, for example, a normal or a lognormal distribution. Such an approach can be accurate for fitting the "body" of the distribution, but will be grossly inaccurate in the tail of the distribution: the skewness of the actual distribution or the heaviness of its tail will be difficult to match. As a result, any prediction of the statistics of rare events, lying far in the tail, will be very inaccurate.

Statistical Blockade

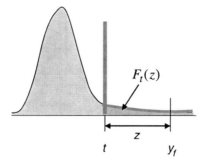

Figure 3.1. A possible skewed distribution for some SRAM metric (e.g., write time)

Let F denote the cumulative distribution function (CDF) of the write time y, and let us define a tail threshold t to mark the beginning of the tail (e.g., the 99-th percentile). Let z be the excess over the threshold t. We can write the *conditional* CDF of the tail as

$$F_t(z) = P(Y - t \leq z | Y > t) = \frac{F(z+t) - F(t)}{1 - F(t)}, \qquad (3.1)$$

and the overall CDF as

$$F(z+t) = (1 - F(t))F_t(z) + F(t). \qquad (3.2)$$

If we know $F(t)$ and can estimate the conditional CDF of the tail $F_t(z)$ accurately, we can accurately estimate rare event statistics. For example, the yield for some extreme threshold $y_f > t$ is given as

$$F(y_f) = (1 - F(t))F_t(y_f - t) + F(t), \qquad (3.3)$$

and the corresponding failure probability $\bar{F}(y_f) = 1 - F(y_f)$ is given as

$$\bar{F}(y_f) = (1 - F(t))(1 - F_t(y_f - t)). \qquad (3.4)$$

$F(t)$ can be accurately estimated using a few thousand simulations, since t is not too far out in the tail. Then, the problem here is to efficiently estimate the conditional tail CDF F_t as a simple analytical form, which can then be used to compute statistical metrics such as (3.3) and (3.4) for rare events. Of course, here we assume that any threshold y_f of interest will be far into the tail, such that $y_f \gg t$. This is easily satisfied for any real HRC scenario, for example our 1 Mb cache example from Sect. 3.1. We also assume that the extreme values of interest lie only in the *upper* tail of the distribution. This is without any loss of generality, because any lower tail can be converted to the upper tail by replacing $y = -y$, and if both tails are of interest (with symmetrical tail thresholds), we can use $y = |y|$. This same approach of fitting a CDF to the exceedances over some threshold has been developed and widely applied by hydrologists under the name of the *peaks over threshold* (POT) method [EKM03]. In their case though, the data is from historical record and not synthetically generated. We now look at some results from extreme value theory that are directly applicable to the problem of estimating the tail CDF.

3.2.2 Extreme Value Theory: Tail Distributions

Suppose that $Y_1, Y_2, \ldots$ is a sequence of independent, identically distributed random variables from the CDF F. For any sample $\{Y_1, Y_2, \ldots, Y_N\}$ of size N, define the *sample maximum* as

$$M_N = \max(Y_1, Y_2, \ldots, Y_N), \quad N \geq 2. \tag{3.5}$$

The probability of $M_N \leq y$ is the probability of all of $\{Y_1, Y_2, \ldots, Y_N\}$ being $\leq y$. Hence,

$$P(M_N \leq y) = P(Y_1 \leq y, \ldots, Y_N \leq y) = \prod_{i=1}^{N} P(Y_i \leq y) = F^N(y). \tag{3.6}$$

An important result from EVT addresses the question: *What are the possible limiting distributions of M_N as $N \to \infty$?* This result is stated in the following theorem by Fisher and Tippett [FT28].

THEOREM 3.1 (Fisher–Tippett [FT28]). *If there exist normalizing constants a_N, b_N, and some non-degenerate CDF H, such that*

$$P\left(\frac{M_N - b_N}{a_N} \leq y\right) = F^N(a_N y + b_N) \to H(y) \quad \text{as } N \to \infty, \; y \in \mathbb{R}, \tag{3.7}$$

then H belongs to the type of one of the following three CDFs:

$$\Phi_\alpha(y) = \begin{cases} 0, & y \leq 0 \\ e^{-y^{-\alpha}}, & y > 0 \end{cases}, \quad \alpha > 0 \; (\text{Fréchet}), \tag{3.8}$$

$$\Psi_\alpha(y) = \begin{cases} e^{-(-y)^\alpha}, & y \leq 0 \\ 1, & y > 0 \end{cases}, \quad \alpha > 0 \; (\text{Weibull}), \tag{3.9}$$

$$\Lambda(y) = e^{-e^{-y}}, \quad y \in \mathbb{R} \; (\text{Gumbel}). \tag{3.10}$$

This amazing result formed the foundation of estimation of rare event statistics. Roughly, it says that for a very large class of CDFs, we can model the distribution of the normalized sample maximum M_N as one of three standard distributions: Fréchet, Weibull and Gumbel. These three CDFs can be combined together into a *generalized extreme value* (GEV) distribution:

$$H_\xi(y) = \begin{cases} e^{-(1-\xi y)^{1/\xi}}, & \xi \neq 0 \\ e^{-e^{-y}}, & \xi = 0 \end{cases}, \quad \text{where } 1 - \xi y > 0. \quad (3.11)$$

The three CDFs are obtained as follows.

- $\xi = -\alpha^{-1} < 0$ gives the Fréchet CDF Φ_α,
- $\xi = \alpha^{-1} > 0$ gives the Weibull CDF Ψ_α, and
- $\xi = 0$ gives the Gumbel CDF Λ.

The condition (3.7) is commonly stated as F *lies in the maximum domain of attraction of* H, or $F \in \text{MDA}(H)$. Hence, for non-degenerate H, Theorem 3.1 can be stated succinctly as

$$F \in \text{MDA}(H) \quad \Rightarrow \quad H \text{ is of type } H_\xi.$$

It is interesting to note the similarity between this theorem regarding maxima and the popular central limit theorem (CLT), which provides the limiting distribution for the *sum* of i.i.d. random variables. The most popular form of the CLT is as follows.

THEOREM 3.2 (Central Limit Theorem). *Define* $S_N = Y_1 + Y_2 + \cdots + Y_N$ *as the sample sum of* N *i.i.d. random variables from some CDF* F. *Let* $\mu = E(Y)$ *be the mean, and* $\sigma^2 = E[(y - \mu)^2]$ *be the variance for* F. *If* $\sigma < \infty$, *then*

$$P\left(\frac{S_N - \mu N}{\sigma \sqrt{N}} \leq y\right) \to \Phi(y) \quad \text{as } N \to \infty, \ y \in \mathbb{R}, \quad (3.12)$$

where Φ *is the standard normal CDF with mean 0 and variance 1.*

Φ, the standard normal CDF, is not to be confused with Φ_α, the Fréchet CDF. We use this potentially confusing notation for consistency with standard literature. Comparing (3.12) and (3.7), we easily see the parallels between the two theorems, made explicit in Table 3.1. Of course, the limiting distribution for maxima is more complex than that for sums (of RVs with *finite* variance) because it has an extra parameter ξ, and the normalizing constants have a more complex dependence on the CDF F. [EKM03] provides these constants for some common distribution types of F. Also, for a general form of the CLT that handles infinite variance, see [EKM03].

CLT	S_N	$\sigma\sqrt{N}$	μN	Φ standard normal
Fisher–Tippett	M_N	a_N	b_N	H_ξ GEV

Table 3.1. The Fisher–Tippett theorem for maxima is congruent to the central limit theorem for sums

H_ξ	Distributions in MDA(H_ξ)	Expression
$\Phi_{-1/\xi}$	Cauchy	$F(y) = \frac{1}{2} + \frac{\arctan(y)}{\pi}$
	Pareto	$F(y) = 1 - y^{-\alpha}$, $K > 0$, $\alpha = -1/\xi > 0$
	Loggamma	$f(y) = \frac{\alpha^\beta}{\Gamma(\beta)}(\ln y)^{\beta-1} y^{-\alpha-1}$, $y > 1$, $\alpha, \beta > 0$
$\Psi_{1/\xi}$	Uniform	$f(y) = 1$, $y \in (0,1)$
	Beta	$f(y) = \frac{\Gamma(a+b)}{\Gamma(a)\Gamma(b)} y^{a-1}(1-y)^{b-1}$, $y \in (0,1)$, $a,b, > 0$
Λ	Normal	$f(y) = \frac{e^{-(y-\mu)^2/2\sigma^2}}{\sqrt{2\pi}\sigma}$
	Lognormal	$f(y) = \frac{e^{-(\ln y-\mu)^2/2\sigma^2}}{\sqrt{2\pi}\sigma y}$, $y > 0$, $\mu \in \mathbb{R}$, $\sigma > 0$
	Gamma	$f(y) = \frac{\beta^\alpha}{\Gamma(\alpha)} y^{\alpha-1} e^{-\beta y}$, $y > 0$, $\alpha, \beta > 0$
	Exponential	$F(x) = 1 - e^{-\lambda y}$, $y > 0$, $K, \lambda > 0$

Table 3.2. Some common distributions lying in MDA(H_ξ). For a longer list, see [EKM03]. F denote the CDF and f the PDF

The conditions for which $F \in \text{MDA}(H)$ for some non-degenerate H, although tighter than for the general CLT, are quite general for most practical purposes, and known well. Gnedenko [Gne43] provided the first rigorous proof for the Fisher–Tippett theorem, showing conditions on F required for the convergence to each of the three limiting CDFs. We state the conditions in Sect. 3.2.3. For now, we only list some common distributions belonging to MDA(H_ξ), in Table 3.2, and immediately proceed to the result due to Balkema and de Haan [BdH74] and Pickands [Pic75], that forms the basis for our proposed tail modeling method. We recall the definition of F_t as the conditional tail CDF for a tail threshold t, as in (3.1). Then, the following is true.

THEOREM 3.3 (Balkema and de Haan [BdH74], and Pickands [Pic75]). *For every $\xi \in \mathbb{R}$, $F \in \text{MDA}(H_\xi)$ if and only if*

$$\lim_{t \to \infty} \sup_{z \geq 0} |F_t(z) - G_{\xi,\beta(t)}(z)| = 0 \qquad (3.13)$$

for some positive function $\beta(t)$, where $G_{\xi,\beta}(z)$ is the generalized Pareto distribution *(GPD)*

$$G_{\xi,\beta}(z) = \begin{cases} 1 - (1 - \xi\frac{z}{\beta})^{1/\xi}, & \xi \neq 0,\ z \in D(\xi,\beta) \\ 1 - e^{-z/\beta}, & \xi = 0,\ z \geq 0 \end{cases}, \qquad (3.14)$$

where

$$D(\xi,\beta) = \begin{cases} [0,\infty), & \xi \leq 0 \\ [0,\beta/\xi], & \xi > 0 \end{cases}.$$

In other words, for any distribution F in the maximum domain of attraction of the GEV distribution, the conditional tail distribution F_t converges to a GPD as we move further out in the tail.

This is an extremely useful result: it implies that, if we can generate enough points in the tail of a distribution $(y \geq t)$, in most practical cases, we can fit the simple, analytical GPD to the data and make predictions further out in the tail. This approach would be independent of the circuit or the performance metric being considered. Of course, two important questions remain:

1. How do we efficiently generate a large number of points in the tail $(y \geq t)$?

2. How do we fit the GPD to the generated tail points?

For answers to these questions, the reader may jump forward to Sect. 3.2.4. We now review the conditions on F for these EVT limit theorems to hold.

3.2.3 Tail Regularity Conditions Required for $F \in \text{MDA}(H_\xi)$

Until now we have "hand-waved" our way through the EVT limit theorems by saying that they apply to "large classes" of CDFs F. In this section, we review the concrete necessary and sufficient conditions on the tail of F, that completely characterize the maximum domain of attraction of the GEV H_ξ (MDA(H_ξ)). Sufficient conditions were provided by von Mises [vM36], and Gnedenko first derived the complete MDA(H_ξ) in [Gne43]. We review the characterization of MDA(H_ξ) using the presentations by Gnedenko, and in [EKM03].

From the form of the GPD in (3.14) and Fig. 3.2, intuition tells us that the tail of F should show some "smoothness" or "regularity" in its variation as we move farther out in the tail. Karamata's mathematical definition of regularity [Kar33] is relevant here:

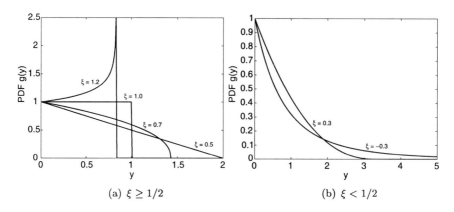

Figure 3.2. The probability density function for a GPD with $\beta = 1$. We get long unbounded tails for $\xi \leq 0$

1) A positive, integrable function $g(y)$ on $(0, \infty)$ is *slowly varying* at ∞ if
$$\lim_{y \to \infty} \frac{g(ky)}{g(y)} = 1, \quad \forall k > 0. \tag{3.15}$$
We write this as $g \in \mathcal{R}_0$.

2) A positive, integrable function $f(y)$ on $(0, \infty)$ has *regular variation* of index α at ∞ if
$$\lim_{y \to \infty} \frac{f(ky)}{f(y)} = t^\alpha, \quad \forall k > 0. \tag{3.16}$$
We write this as $g \in \mathcal{R}_\alpha$. Note that a regularly varying function $f \in \mathcal{R}_\alpha$ can be written as
$$f(y) = y^\alpha g(y), \tag{3.17}$$
where $g(y)$ is a slowly varying function ($g \in \mathcal{R}_0$). Some examples of functions regularly varying at ∞ are
$$y^\alpha, \quad y^\alpha \ln(1+y), \quad (y \ln(1+y))^\alpha,$$
for any real α, while the following are not regularly varying:
$$1 + \cos(y), \quad e^{[\ln(1+y)]},$$
where $[\cdot]$ gives the integer part.

We also define y_F as the upper or right endpoint of F, such that
$$F(y_F) = 1, \quad \text{and} \quad F(y) < 1 \text{ for } y < y_F. \tag{3.18}$$

For F with infinite support, $y_F = \infty$. Now we can state the characterizations of $\text{MDA}(H_\xi)$.

Theorem 3.4 ($\text{MDA}(H_\xi)$ for $\xi < 0$; i.e., $\text{MDA}(\Phi_\alpha)$ for $\alpha = -\xi^{-1}$). *CDF $F \in \text{MDA}(\Phi_\alpha)$, $\alpha > 0$, if and only if*

$$1 - F(y) \in \mathcal{R}_{-\alpha}. \tag{3.19}$$

Theorem 3.5 ($\text{MDA}(H_\xi)$ for $\xi = 0$; i.e., $\text{MDA}(\Lambda)$). *CDF $F \in \text{MDA}(\Lambda)$, if and only if there exists some continuous function $a(y)$, such that $\lim_{y \uparrow y_F} a(y) = 0$, and*

$$\lim_{y \uparrow y_F} \frac{1 - F(y(1 + a(y)t))}{1 - F(y)} = e^{-t}. \tag{3.20}$$

Here $y \uparrow y_F$ means convergence from the left; i.e., $y \not> y_F$.

Theorem 3.6 ($\text{MDA}(H_\xi)$ for $\xi > 0$; i.e., $\text{MDA}(\Psi_\alpha)$ for $\alpha = \xi^{-1}$). *CDF $F \in \text{MDA}(\Psi_\alpha)$, $\alpha > 0$, if and only if*

$$1 - F\left(y_F - \frac{1}{y}\right) \in \mathcal{R}_{-\alpha}. \tag{3.21}$$

Detailed proofs and discussion of these conditions can be found in [Res87]. Most common continuous CDFs satisfy one of these conditions and, hence, allow a GPD approximation for their tails, as per Theorem 3.3. Examples of common *discrete* distributions that do not satisfy these conditions are the Poisson and hypergeometric distributions, as discussed in [EKM03]. For our problems of statistical analysis of high replication circuits, however, we expect mainly to see continuous CDFs with long smooth tails, allowing reasonable application of the GPD approximation. This argument is supported by the promising results in Sects. 3.3.4 and 3.4.4.

3.2.4 Estimating the Tail: Fitting the GPD to Data

For now, let us suppose that we can generate a reasonably large number of points in the tail of our performance distribution. For this we might, theoretically, use standard Monte Carlo simulation with an extremely large sample size, or, more practically, the statistical blockade sampling method proposed in Sect. 3.3.3. Let this data be $\mathbf{Z} = (Z_1, \ldots, Z_n)$, where each Z_i is the *exceedance* over the tail threshold t ($Z_i > 0$, $\forall i$). All Z_i are i.i.d. random variables with common CDF F_t. Then we have the problem of estimating the optimal GPD parameters ξ, β from this tail data, so as to best fit the conditional tail CDF F_t. There are several options; we review three of the most popular ones here. In particular we focus on methods that require no manual effort and can be completely optimized. For manual methods based on graphical exploration of the data, see [EKM03].

3.2.4.1 Maximum Likelihood Estimation

Maximum likelihood estimation (MLE) is a standard statistical estimation technique that tries to estimate those model parameters (here ξ, β of the GPD) that maximize the "chances" of obtaining the data that we have observed. The probability density function of a GPD $G_{\xi,\beta}$ is given as

$$g_{\xi,\beta}(z) = \begin{cases} \frac{1}{\beta}(1 - \xi\frac{z}{\beta})^{1/\xi - 1}, & \xi \neq 0, \ z \in D(\xi,\beta) \\ \frac{1}{\beta}e^{-z/\beta}, & \xi = 0, \ z \geq 0 \end{cases}, \quad (3.22)$$

where $D(\xi,\beta)$ is defined in Theorem 3.3. Recall that all Z_i are i.i.d. random variables with common CDF F_t. We assume that F_t is of the form of a GPD. The *likelihood* ("chances") of having seen this data from an underlying GPD is the multivariate probability density associated with it, and is given as

$$\mathcal{L}(\xi,\beta|\mathbf{Z}) = g_{\xi,\beta}(Z_1, \ldots, Z_n) = \prod_{i=1}^{n} g_{\xi,\beta}(Z_i). \quad (3.23)$$

Since $\mathcal{L}(\xi,\beta|\mathbf{Z})$ can be too small for accurate computation with finite accuracy, it is typical to use the *log-likelihood function*

$$\ell(\xi,\beta|\mathbf{Z}) = \ln(\mathcal{L}(\xi,\beta|\mathbf{Z})) = \sum_{i=1}^{n} \ln(g_{\xi,\beta}(Z_i)), \quad (3.24)$$

which increases monotonically with $\mathcal{L}$. MLE then computes (ξ,β) to maximize this log-likelihood, as

$$(\hat{\xi},\hat{\beta})_{\text{mle}} = \arg\max_{\xi,\beta} \sum_{i=1}^{n} \ln(g_{\xi,\beta}(Z_i)). \quad (3.25)$$

Substitution of (3.22) in (3.25) and subsequent algebra allows for a simplification to a one dimensional search, that can be exploited by a careful implementation of a Newton–Raphson algorithm, as shown in [Gri93].

Smith [Smi87] studies convergence when F_t is not exactly of GPD form, and provides limit results for the distributions of $(\hat{\xi},\hat{\beta})_{\text{mle}}$ for each of the three cases, $F \in \text{MDA}(\Phi_{-1/\xi})$, $F \in \text{MDA}(\Lambda)$ and $F \in \text{MDA}(\Psi_{1/\xi})$. For $\xi < \frac{1}{2}$, the MLE estimates are asymptotically normal and efficient (bias = 0) under certain regularity assumptions on F. If (ξ, β) are the exact values to be estimated, then as the sample size $n \to \infty$, the variance of the MLE estimates is given as

$$\text{var}\begin{bmatrix}\hat{\xi}\\\hat{\beta}\end{bmatrix} \to \frac{1-\xi}{n}\begin{bmatrix}1-\xi & \beta\\ \beta & 2\beta^2\end{bmatrix}, \quad \xi < \frac{1}{2}. \quad (3.26)$$

When $\xi \geq \frac{1}{2}$, MLE convergence can be difficult and special techniques are needed [Smi85]. However, $\xi \geq \frac{1}{2}$ is usually rare, since it corresponds to a finite tail with $g_{\xi,\beta}(z) > 0$ at the endpoint (Fig. 3.2).

3.2.4.2 Moment Matching

An ad hoc way of estimating the GPD parameters is to match the moments of the GPD with the moments of the data, as we now describe. According to [HW87], the p-th moment for the GPD exists if $\xi > -1/p$. In many practical cases we expect finite mean and variance, and hence, existence of the first and second moments. The mean and variance for $G_{\xi,\beta}$ are given as

$$\mu = \frac{\beta}{1+\xi}, \quad \sigma^2 = \frac{\beta^2}{(1+\xi)^2(1+2\xi)}, \quad (3.27)$$

respectively. Equating these with the *sample* mean and variance, $\hat{\mu}$ and $\hat{\sigma}^2$, respectively, we can compute estimates of ξ and β:

$$\hat{\xi} = \frac{1}{2}\left(\frac{\hat{\mu}^2}{\hat{\sigma}^2} - 1\right), \quad \hat{\beta} = \frac{\hat{\mu}}{2}\left(\frac{\hat{\mu}^2}{\hat{\sigma}^2} + 1\right). \quad (3.28)$$

For $\xi > -1/4$, the estimates are asymptotically normal. See [HW87] for variance estimates for this limit distribution, which are skipped here since this method is not as popular or reliable as the other two methods (MLE and PWM).

3.2.4.3 Probability-Weighted Moment Matching

Probability-weighted moments (PWMs) [Hos86] of a continuous random variable Y with CDF F are generalizations of the standard moments, and are defined as

$$M_{p,r,s} = E[Y^p F^r(Y)(1-F(Y))^s]. \quad (3.29)$$

The standard p-th moment is given by $M_{p,0,0}$. For the GPD, we have a convenient relationship between $M_{1,0,s}$ and (ξ, β), given by

$$m_s = M_{1,0,s} = \frac{\beta}{(1+s)(1+s+\xi)}, \quad \xi > 0. \quad (3.30)$$

Then, we can write

$$\beta = \frac{2m_0 m_1}{m_0 - 2m_1}, \qquad \xi = \frac{m_0}{m_0 - 2m_1} - 2. \qquad (3.31)$$

We estimate these PWMs from the data sample, as

$$\hat{m}_s = \frac{1}{n}\sum_{i=1}^{n}(1-q_i)^s Y_{i,n}, \qquad (3.32)$$

where

$$Y_{1,n} \leq Y_{2,n} \leq \cdots \leq Y_{n,n} \qquad (3.33)$$

is the *ordered sample*, and

$$q_i = \frac{i+\gamma}{n+\delta} \qquad (3.34)$$

with $\gamma = -0.35, \delta = 0$, as suggested in [HW87]. The estimates $(\hat{\xi}, \hat{\beta})$ converge to the exact values as $n \to \infty$, and are asymptotically normally distributed with covariance given by

$$\operatorname{var}\begin{bmatrix}\hat{\xi}\\\hat{\beta}\end{bmatrix} \to \frac{n^{-1}}{(1+2\xi)(3+2\xi)}$$
$$\times \begin{bmatrix} (1+\xi)(2+\xi)^2(1+\xi+2\xi^2) & \beta(2+\xi)(2+6\xi+7\xi^2+2\xi^3) \\ \beta(2+\xi)(2+6\xi+7\xi^2+2\xi^3) & \beta^2(7+18\xi+11\xi^2+2\xi^3) \end{bmatrix}.$$
$$(3.35)$$

Based on an extensive simulation study, [HW87] suggests that the PWM method often has lower bias than moment matching and MLE for sample sizes up to 500. Also, the MLE search (3.25) is shown to suffer from some convergence problems when ξ is estimated close to $1/2$. Finally, the study also suggests that PWM matching gives more reliable estimates of the variability of the estimated parameters, as per (3.35). Based on these reasons, we choose PWM matching for the purpose of this thesis.

Once we have estimated a GPD model of the conditional CDF above a threshold t, we can estimate the failure probability for any value y_f by substituting the GPD in (3.4) as

$$P(Y > y_f) \approx (1 - F(t))(1 - G_{\xi,\beta}(y_f - t)). \qquad (3.36)$$

The next section addresses the important remaining question: How do we efficiently generate a large number of points in the tail $(y \geq t)$?

3.3 Statistical Blockade

Before we can introduce the proposed rare event sampling technique, a review of the concept of *classification* from machine learning seems appropriate, given that it will form a cornerstone for the proposed technique. Readers familiar with classifiers may skip to Sect. 3.3.3.

3.3.1 Classification

Consider the problem of detecting spam in incoming email. A spam detector is some computer program that takes in as inputs, certain features of any new incoming email message, and predicts whether the message is "email" or "spam". The input *features* may be such as the sender email id, the occurrence of words commonly seen in spam email, etc. We can think of this program as a *function* with these features as inputs and a *categorical* or *discrete* output. The output can assume one of two possible values, "email" or "spam", which we call *classes*. Any such function, that predicts the class of any given input vector, is called a *classifier*, and this act of such prediction is called *classification*. In the general case, we may have any number of classes, and any number of input features. Consider, for example, Fig. 3.3, where there are two input features $\mathbf{x} = (x_1, x_2)$ and three possible classes denoted by ○, △ and +. A classifier would be a function $C(\mathbf{x})$ that, given some input vector $\mathbf{x}$, returns 0, 1 or 2 (for ○, △ or +, respectively); i.e., it predicts the class that $\mathbf{x}$ belongs to. Hence, the classifier defines some inter-class boundaries in the input space: Fig. 3.3 also shows a possible set of such boundaries.

For a two-class problem with one *linear* boundary, a simple classifier can be based on linear regression as

$$C(\mathbf{x}) = \text{sign}[\mathbf{x}^T \mathbf{w} + b], \quad (3.37)$$

where $\mathbf{w}$ and b are chosen such that the linear function $\mathbf{x}^T \mathbf{w} + b$ is > 0 for any point in one class and < 0 for any point in the other class. The sign[·] function returns the sign of the linear function, converting its argument from a real variable to a discrete valued variable $\in \{-1, 1\}$. The boundary defined by such a linear regression based classifier is

$$\mathbf{x}^T \mathbf{w} + b = 0. \quad (3.38)$$

Before any classifier can be used, it has to be "trained"; i.e., optimal values for its parameters have to be determined. In our linear regression example, the parameters are the elements of $\mathbf{w}$, and b. Such training starts from a training sample of points for which the class values are known, and then computes the parameters so as to minimize the error

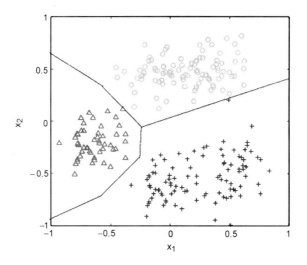

Figure 3.3. Example with two input features (dimensions) and three possible classes. The *solid lines* show a possible set of boundaries dividing the classes

between the classifier predictions and the actual class values for all the training points. Denote the vector of all classifier parameters by **p**. Let $(\mathbf{x}_1, y_1), (\mathbf{x}_2, y_2), \ldots, (\mathbf{x}_n, y_n)$ be the n training points, where y_i is the class of the i-th point $\mathbf{x}_i$. We compute **p** such that

$$\min_{\mathbf{p}} \text{Error}(\{C(\mathbf{x}_i), y_i\}_{i=1}^n), \tag{3.39}$$

where the precise definition of Error depends on the particular classifier and optimization method used. We now review one particularly successful type of classifier, called *support vector machines*, which we use for the implementation of statistical blockade in this thesis.

3.3.2 Support Vector Classifier

A support vector machine (SVM) classifier uses the *optimal separating hyperplane* as the decision boundary between classes in the input space. We provide an introductory discussion of the basic ideas behind SVMs in this section. SVMs enjoy extensive application for statistical inference in a wide variety of problem domains; see [Bur98] for example. There are good reasons for this widespread popularity of SMVs. The basic idea is intuitive and simple, and it allows for classifiers with very good generalizability (low overfitting), relative to many other competing approaches [Bur98][HTF01]. [Bur98] provides a good tutorial of SVMs in the classification context.

The basic SVM separates two classes with a linear boundary (the optimal separating hyperplane), although, it has been generalized to easily

Statistical Blockade

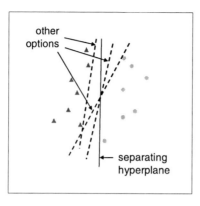

Figure 3.4. Example of data separable with a hyperplane. In this case, the hyperplane is a *straight line*. Multiple options for the separating hyperplane are shown as *dashed straight lines*

handle nonlinear boundaries and multiple (≥ 3) classes, as discussed in [Bur98] and [HTF01], for instance. Here we restrict our discussion to the two-class, linear classifier that we use for our implementation of the proposed statistical blockade method. First, let us assume that the training points are separable with a linear boundary: there exists a hyperplane that can completely divide the two classes without any errors. This case is shown in Fig. 3.4. We call this hyperplane, the *separating hyperplane*. We start with a linear classification rule, as in (3.37)

$$C(\mathbf{x}) = \text{sign}[\mathbf{x}^T \mathbf{w} + b] \qquad (3.40)$$

with the separating hyperplane given by

$$S : \mathbf{x}^T \mathbf{w} + b = 0. \qquad (3.41)$$

Recall that the training data consists of n pairs $(\mathbf{x}_1, y_1), (\mathbf{x}_2, y_2), \ldots, (\mathbf{x}_n, y_n)$, where $y_1 \in \{-1, 1\}$. If we can correctly orient this separating hyperplane S; i.e., there are no training errors, then we expect that

$$\begin{aligned} \mathbf{x}_i^T \mathbf{w} + b > 0 & \quad \text{whenever } y_i = 1, \quad \text{and} \\ \mathbf{x}_i^T \mathbf{w} + b < 0 & \quad \text{whenever } y_i = -1. \end{aligned} \qquad (3.42)$$

Then,

$$y_i(\mathbf{x}_i^T \mathbf{w} + b) > 0, \quad \forall i. \qquad (3.43)$$

Even with this condition, there are several choices for the separating hyperplane, as shown in Fig. 3.4: which is the best choice? The best choice is called, understandably, the *optimal* separating hyperplane, and it maximizes the distance to the point nearest to it from either class,

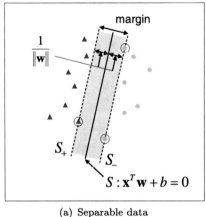
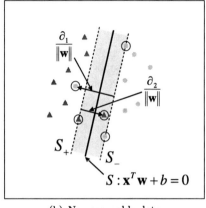

(a) Separable data   (b) Non-separable data

Figure 3.5. Optimal separating hyperplane maximizes the margin between the nearest points from the two classes, shown here for both separable and non-separable data. The support vectors lie on the margin hyperplanes, and are *circled* in both cases

as shown in Fig. 3.5(a). For any such hyperplane, let d_+ and d_- denote the distance to the nearest point from classes 1 and -1, respectively. We define the *margin* of the separating hyperplane as $d_+ + d_-$. Suppose, we always center the hyperplane in the margin, so that $y_i(\mathbf{x}_i^T \mathbf{w} + b) = 1$ for the nearest points in either class. The signed perpendicular distance between any point $\mathbf{x}$ and the hyperplane is given by

$$\frac{\mathbf{x}^T \mathbf{w} + b}{\|\mathbf{w}\|}. \tag{3.44}$$

Hence,

$$d_+ = d_- = \frac{1}{\|\mathbf{w}\|}, \tag{3.45}$$

giving us a margin of $\frac{2}{\|\mathbf{w}\|}$. The optimal separating hyperplane is obtained by maximizing this margin as

$$\min_{\mathbf{w},b} \frac{1}{2}\|\mathbf{w}\|^2 \tag{3.46}$$
$$\text{subject to} \quad y_i(\mathbf{x}_i^T \mathbf{w} + b) \geq 1, \ i = 1, \ldots, n.$$

The Lagrangian formulation of this optimization problem is given as

$$\min_{\mathbf{w},b} \frac{1}{2}\|\mathbf{w}\|^2 - \sum_{i=1}^{n} \mu_i [y_i(\mathbf{x}_i^T \mathbf{w} + b) - 1], \tag{3.47}$$

where μ_i are the Lagrangian multipliers. At the solution, the derivatives of the objective are zero, giving us

$$\mathbf{w} = \sum_{i=1}^{n} \mu_i y_i \mathbf{x}_i. \qquad (3.48)$$

The multipliers μ_i are nonzero only for those points that satisfy the *equality* in the constraints in (3.46); i.e., for points exactly on the margin hyperplanes (S_+, S_-). Hence, the orientation of the separating hyperplane is determined only by those points that lie on the margin hyperplanes, giving them the name of *support points*. Figure 3.5(a) shows the support points as circled.

All of this development assumed that the training points are separable using a hyperplane. Of course, in many practical situations this is not true: even the optimal separating hyperplane can suffer from misclassifications due to "overlap" in the classes. Figure 3.5(b) shows an example. The complete linear SVM formulation accounts for these cases by including positive "slack" variables $\delta_i, i = 1, \ldots, n$ in the constraints of (3.46) as

$$y_i(\mathbf{x}_i^T \mathbf{w} + b) \geq 1 - \delta_i, \quad i = 1, \ldots, n,$$
$$\delta_i \geq 0, \quad i = 1, \ldots, n. \qquad (3.49)$$

For an error to occur on point i, $\delta_i > 1$, hence $\sum_{i=1}^{n} \delta_i$ is an upper bound on the number of misclassifications. Since we wish to minimize the training error *and* maximize the margin size, a natural choice for the optimization problem is

$$\min_{\mathbf{w},b} \frac{1}{2}\|\mathbf{w}\|^2 + \gamma \sum_{i=1}^{n} \delta_i \quad \text{subject to (3.49)}, \qquad (3.50)$$

where γ is a user-supplied tuning parameter. Note that the distance of a misclassified point from its margin hyperplane is $\frac{\delta_i}{\|\mathbf{w}\|}$, as shown in Fig. 3.5(b). All points on their margin hyperplane or on the wrong side of it are called the *support vectors*, since they alone determine the estimates of $\mathbf{w}$ and b, in a manner similar to the separable case. This optimization problem is easiest to solve in its dual form. We refer the reader to [Bur98] and [HTF01] for further details on these aspects, and generalizations of SVMs to nonlinear boundaries and multiple classes. SVMs are a popular, well researched classification strategy, and optimized software implementations are readily available; for example, SVM^{light} [Joa99] and WEKA [WF05].

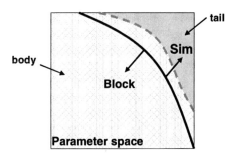

Figure 3.6. The tail and body regions in the statistical parameter space. The *dashed line* is the exact tail region boundary for tail threshold t. The *solid line* is the relaxed boundary modeled by the classifier for a classification threshold $t_c < t$

3.3.3 The Statistical Blockade Algorithm

We are now almost ready to synthesize all the pieces of the proposed statistical blockade: only a mapping of the theory in the foregoing sections to various aspects of the high replication circuit problem is needed. As in Chap. 1, we let any circuit performance metric, or simply, output y be computed as

$$y = f_{\text{sim}}(\mathbf{x}). \tag{3.51}$$

Here, $\mathbf{x}$ is a point in the statistical parameter (e.g., V_t, t_{ox}) space, or simply, the input space, and f_{sim} includes expensive SPICE simulation. We assume that y has some probability distribution F, with an extended tail. Suppose, we define a large tail threshold t for y, then from the developments in Sect. 3.2.2 we know that we can approximate the conditional tail CDF F_t by a generalized Pareto distribution $G_{\xi,\beta}$. Section 3.2.4 shows how we can estimate the GPD parameters (ξ, β) from data drawn from the tail distribution. We now make explicit our efficient tail sampling strategy that will generate the tail points for fitting this GPD.

Corresponding to the tail of output distribution, we expect a "tail region" in the input space: any statistical parameter values drawn from this tail region will give an output value $y > t$. Figure 3.6 shows an example of such a tail region for two inputs. The rest of the input space is called the "body" region, corresponding to the body of the output distribution F. In Fig. 3.6 these two regions are separated by a dashed line. The key idea behind the proposed sampling technique is to identify the tail region and simulate only those Monte Carlo points that are likely to lie in this tail region. Here, we exploit the common fact that *generating* the random values for a Monte Carlo sample point is very cheap compared to actually *simulating* the point as in (3.51). Hence, if

Statistical Blockade 143

Algorithm 3.1 The *statistical blockade* algorithm for efficiently sampling rare events and estimating their probability distribution

Require: training sample size n_0 (e.g., 1,000); total sample size n; percentages p_t (e.g., 99%), p_c (e.g., 97%)
1: $\mathbf{X} = \text{MonteCarlo}(n_0)$
2: $\mathbf{y} = f_{\text{sim}}(\mathbf{X})$
3: $t = \text{Percentile}(\mathbf{y}, p_t)$
4: $t_c = \text{Percentile}(\mathbf{y}, p_c)$
5: $C = \text{BuildClassifier}(\mathbf{X}, \mathbf{y}, t_c)$ // C is a classifier
6: $\mathbf{y} = f_{\text{sim}}(\text{Filter}(C, \text{MonteCarlo}(n)))$
7: $\mathbf{y}_{\text{tail}} = \{y_i \in \mathbf{y} : y_i > t\}$
8: $(\xi, \beta) = \text{FitGPD}(\mathbf{y}_{\text{tail}} - t)$

we generate points as in standard Monte Carlo, but *block* – not simulate – those points that are unlikely to fall in the tail region, we can drastically cut down the total time spent. This reduction in time spent is drastic because we are trying to simulate only the rare events, which by definition constitute a very small percentage of the total Monte Carlo sample size. As might be obvious to the discerning reader, we can use a classifier to distinguish the tail and body regions, and to block out the body points. For any candidate point in the input space, generated from standard Monte Carlo, the classifier can predict its membership in either the "body" or the "tail" classes. Only the "tail" points are then simulated.

To build this model of the tail region boundary, the classifier can be trained with a small (e.g., 1,000 points) training set of simulated Monte Carlo sample points. However, it is difficult, if not impossible to build an *exact* model of the boundary in general. Misclassifications, at least on points unseen during training, is unavoidable. Hence, we relax the accuracy requirement to allow for classification error. This is done by building the classification boundary at a *classification threshold* t_c that is less than the tail threshold t. Since we have assumed that only the upper (right) tail is relevant, the tail region corresponding to t will be a subset of the tail region corresponding to t_c, if $t_c < t$. This will help to ensure that, even if the classifier is imperfect, it is unlikely that it will misclassify points in the true tail region (for t). The relaxed boundary corresponding to such a t_c is shown as the solid line in Fig. 3.6.

The statistical blockade algorithm is then as in Algorithm 3.1. The algorithm derives its name from the blocking activity of the classifier. We also refer to this classifier as the *blockade filter* and its blocking activity as *blockade filtering*. The thresholds $t = p_t$-th percentile and $t_c = p_c$-th

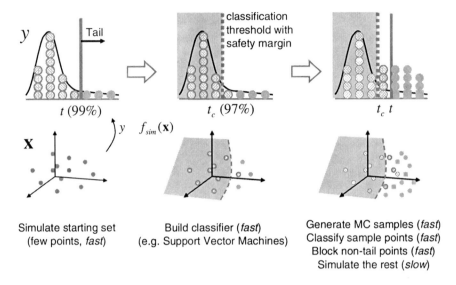

Figure 3.7. The efficient tail (rare event) sampling method of statistical blockade

percentile are estimated from the small initial Monte Carlo run, which also gives the n_0 training points for the classifier. Typical values for these constants are shown in Algorithm 3.1. The function MonteCarlo(n) generates n points in the statistical parameter space, which are stored in the $n \times s$ matrix $\mathbf{X}$, where s is the input dimensionality. Each row of $\mathbf{X}$ is a point in s dimensions. $\mathbf{y}$ is a vector of output values computed from simulations. The function BuildClassifier($\mathbf{X}$, $\mathbf{y}$, t_c) trains and returns a classifier using the training set $(\mathbf{X},\mathbf{y})$ and classification threshold t_c. The function Filter(C, $\mathbf{X}$) blocks the points in $\mathbf{X}$ classified as "body" by the classifier C, and returns only the points classified as "tail". FitGPD($\mathbf{y}_{\text{tail}} - t$) computes the parameters (ξ, β) for the best GPD approximation $G_{\xi,\beta}$ to the conditional CDF of the exceedances of the tail points in $\mathbf{y}_{\text{tail}}$ over t. We can then use this GPD model to compute statistical metrics for rare events, for example, the failure probability for some threshold y_f, as in (3.36). This sampling procedure is also illustrated in Fig. 3.7.

3.3.3.1 Note on Choosing and Unbiasing the Classifier

The algorithm places no restrictions on the choice of classifier. In this thesis, we use support vector machines, described in Sect. 3.3.2. We make some practical observations here that are relevant for the choice of classifier. High replication circuits naturally tend to be small, relatively simple circuits. It is highly unlikely that a complex, large circuit will be replicated thousands to millions of times on the same chip. This level of replication often naturally coincides with simple functionality. As a result, we often do not expect to see drastically nonlinear boundaries for the tail regions of these circuits. Nor do we expect to see very complex topologies of the tail regions. These considerations, along with the safety margin awarded by a classification threshold t_c less than t, led us to use linear SVMs. Indeed, linear SVMs suffer minimally from overfitting issues and from the complex parameter selection problems of nonlinear, kernel-based SVMs. As we shall demonstrate with experiments in Sects. 3.3.4 and 3.4.4, this choice does result in an effective implementation of statistical blockade. For cases where a strongly nonlinear boundary exists, a linear classifier may not suffice, and more sophisticated classification techniques may be required [HTF01]. The statistical blockade framework however, should not need any fundamental change.

An important technical point to note about the classifier construction is as follows. The training set will typically have many more body points than tail points. Hence, even if all or most of the tail points are misclassified, the training error will be low as long as most of the body points are correctly classified. This will result in a classifier that is biased to allow more misclassifications of points in the tail region. However, we need to minimize misclassification of tail points to avoid distorting the statistics of the simulated tail points. Hence, we need to reverse bias the classification error. Using the technique proposed in [MBJ99], we penalize misclassifications of tail points more than misclassifications of body points. In the context of SVMs, let us rewrite the training objective function (3.50) as

$$\min_{\mathbf{w},b} \frac{1}{2}\|\mathbf{w}\|^2 + \gamma_+ \sum_{i:y_i=1} \delta_i + \gamma_- \sum_{i:y_i=-1} \delta_i \quad \text{subject to (3.48)}, \qquad (3.52)$$

where γ_+ and γ_- are possibly different penalty factors for the two classes ("tail" and "body" in our case). If, as in [MBJ99], we choose

$$\frac{\gamma_+}{\gamma_-} = \frac{\text{Number of '}-\text{' training points}}{\text{Number of '}+\text{' training points}} = \frac{\text{Number of "body" points}}{\text{Number of "tail" points}}, \qquad (3.53)$$

we can obtain an unbiased classifier. Any other choice of classifier (instead of SVMs) will also require such asymmetric penalties during training.

3.3.4 Experimental Results

We now apply the statistical blockade method to three test cases:

1) a 6T SRAM cell,

2) a complete 64-bit SRAM column with write driver, and

3) a master–slave flip-flop with the scan chain component.

The initial training sample used to construct each blockade filter is from a standard Monte Carlo run of $n_0 = 1{,}000$ points. The filter is an SVM classifier built using the 97-th percentile of each relevant performance metric as the classification threshold t_c. The tail threshold is defined as the 99-th percentile.

In all cases the rare event statistical metric we compute is the failure probability $\bar{F}(y_f)$ for any failure threshold y_f, using the GPD fit to the tail defined by the tail threshold t. We represent this failure probability as the equivalent quantile y_σ on the standard normal distribution:

$$y_\sigma = \Phi^{-1}(1 - \bar{F}(y_f)) = \Phi^{-1}(F(y_f)) \qquad (3.54)$$

where Φ is the standard normal CDF. For example, a failure probability of $\bar{F} = 0.00135$ implies a cumulative probability of $F = 1 - \bar{F} = 0.99865$. The equivalent point on a standard normal, having the same cumulative probability, is $y_\sigma = 3$. In other words, any y_f with a failure probability of 0.00135 is a "3σ" point.

We can compute $\bar{F}(y_f)$, and hence y_σ, in three different ways:

I. *Empirically*: Run a large Monte Carlo run where all points are fully simulated; i.e., with no use of blockade filtering or EVT. Say we use a sample size of n_{MC} (e.g., 1 million), giving us n_{MC} values $y_i, i = 1, \ldots, n_{\text{MC}}$. Then we can empirically compute $\bar{F}(y_f)$ as

$$\bar{F}(y_f) \approx \frac{|\{y_i : y_i > y_f\}|}{n_{\text{MC}}}. \qquad (3.55)$$

Of course, for any $y_f > \max(\{y_1, \ldots, y_{n_{\text{MC}}}\})$, we will get the same estimate of 0 failure probability, and $y_\sigma = \infty$, since there are no points beyond this y_f to give us any information about such rare events. Hence, the prediction power of the empirical method is limited by the Monte Carlo sample size.

II. *Using GPD model, with no blockade filtering*: We can run a full Monte Carlo run with no filtering, as in the empirical estimation case, but then fit a GPD to the points in the tail, defined by the tail threshold t. These are the points $\{y_i : y_i > t\}$. Using this GPD, $G_{\xi,\beta}$, in (3.36), which we reproduce here for convenience,

$$\bar{F}(y_f) = P(Y > y_f) \approx (1 - F(t))(1 - G_{\xi,\beta}(y_f - t)), \qquad (3.56)$$

we can compute the failure probability. $F(t)$ can be estimated empirically with good accuracy. The GPD model extends the prediction

Statistical Blockade

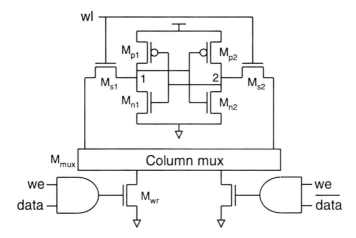

Figure 3.8. A 6-transistor SRAM cell with write driver and column mux

power all the way to ∞. Of course, the confidence in the prediction would decrease as we move to very high values of y_f.

III. *Using statistical blockade*: Here we use the complete statistical blockade flow, where only candidate tail points identified by the blockade filter are simulated, and a GPD tail model is estimated from the actual tail points $y > t$. Here, too, we use (3.56), but the points used to estimate (ξ, β) are obtained from blockade filtering. Further, we use a Monte Carlo sample size that is much smaller than for method II, to test statistical blockade in a practical setting, where we want to use as small a sample size as possible.

For all the test cases we compare the predictions of y_σ from these three methods. Method II gives the most accurate estimates, since it uses a large number of points and no filtering. In some cases, we also show estimates computed using a Gaussian distribution fit to highlight the error in such an approach. Let us now look at the test circuits in more detail, along with the results we obtain.

3.3.4.1 6T SRAM Cell

The first test case is a standard 6T SRAM cell with bit-lines connected to a column multiplexor and a non-restoring write driver, shown in Fig. 3.8. The device and statistical models are the same as for the two-stage opamp in Sect. 1.7.2. We use the Cadence 90 nm Generic PDK library, with independent, normally distributed threshold voltage variation per transistor and a global gate oxide thickness variation, also normally distributed. This gives us a total of 9 statistical parameters. The metric

τ_w (y_f) (FO4)	(I) Standard Monte Carlo	(II) GPD *no* blockade filter	(III) Statistical blockade
2.4	3.404	3.408	3.379
2.5	3.886	3.886	3.868
2.6	4.526	4.354	4.352
2.7	∞	4.821	4.845
2.8	∞	5.297	5.356
2.9	∞	5.789	5.899
3.0	∞	6.310	6.493
Number of simulations	1,000,000	1,000,000	5,379

Table 3.3. Prediction of failure probability as y_σ by methods I, II and III, for a 6T SRAM cell. The number of simulations for statistical blockade includes the 1,000 training samples. The write time values are shown in "fanout of 4" units

being measured is the write time τ_w: the time between the wordline going high, to the non-driven cell node (node 2) transitioning. Here, "going high" and "transitioning" imply crossing 50% of the full voltage change. For methods I and II, we use $n_{\text{MC}} = 1$ million Monte Carlo points. For statistical blockade (method III), 100,000 Monte Carlo points are filtered through the classifier, generating 4,379 tail candidates. On simulating these 4,379 points, 978 true tail points ($\tau_w > t$) were obtained, which were then used to compute a GPD model for the tail conditional CDF. Table 3.3 shows a comparison of the y_σ values estimated by the three different methods. We can see a close match between the predictions by the accurate method II and statistical blockade, method III. Figure 3.9 compares the conditional tail CDFs computed from the empirical method and from statistical blockade, showing a good match.

Some observations highlighting the efficiency of statistical blockade can be made immediately.

- The empirical method fails beyond 2.6 FO4, corresponding to about 1 ppm circuit failure probability, because there are no points generated by the Monte Carlo run so far out in the tail.

- Fitting a GPD model to the tail points (method II) allows us to make predictions far out in the tail, even though we have no points that far out.

- Using blockade filtering, coupled with the GPD tail model, we can drastically reduce the number of simulations (from 1 million to 5,379) with very small change to the tail model.

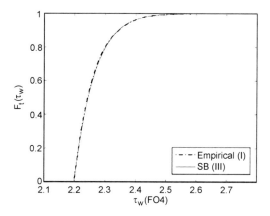

Figure 3.9. Comparison of GPD tail model from statistical blockade (5,379 simulations) and the empirical tail CDF (1 million simulations) for the write time of the 6T SRAM cell

Of course, the tail model cannot be relied on too far out from the available data, as suggested by the increased discrepancy between methods II and III for the largest τ_w values. We further discuss and attack this problem in Sect. 3.4.

3.3.4.2 64-Bit SRAM Column

The next test case is a 64-bit SRAM column, with a non-restoring write driver and column multiplexor, shown in Fig. 3.10. Only one cell is being accessed, while all the other wordlines are turned off. Random threshold variation on all 402 transistors (including the write driver and column mux) are considered, along with a global gate oxide variation. The device and variation models are the same 90 nm technology as for the 2-stage opamp in Sect. 1.7.2. In scaled technologies, leakage current is no longer negligible [RSBS04]. Hence, process variations on transistors that are meant to be inaccessible (or off) can also impact the overall behavior of a circuit. This test case allows us to see the impact of variations in the leakage current passing through the 63 off cells, along with variations in the write driver. Since the BSIM3v3 models [LJC+88] are used, the gate leakage is not well modeled, but the drain leakage is.

Once again we measure the write time, in this case from the wordline wl_0 to node 2, for falling node 2. The number of statistical parameters is 403. Building a reliable classifier with only $n_0 = 1,000$ points in 403 dimensional space is nearly impossible. However, we can reduce the dimensionality by choosing only those dimensions (statistical parameters) that have a significant impact on the write time. We address essentially

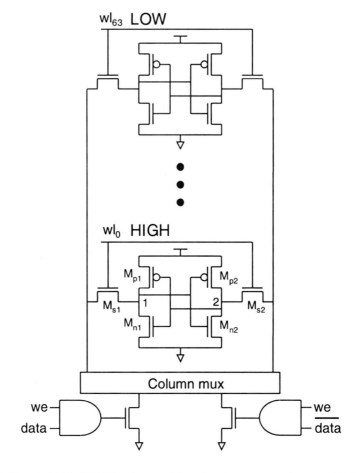

Figure 3.10. A 64-bit SRAM column with write driver and column multiplexor

the same problem – detecting the "important" variables – in both of the other two chapters of this thesis; in particular, see Sects. 1.6.4.1 and 2.5.1. As in these sections, we use Spearman's rank correlation coefficient, ρ_S (1.69), between each statistical parameter and the circuit performance metric to quantitatively estimate the strength of their relationship. For classification, only parameters with $|\rho_S| > 0.1$ are used, reducing the dimensionality to only 11. Figure 3.11(a) shows the sorted magnitudes of the 403 rank correlation values: we can see that only a handful of the statistical parameters have significant correlation with the write time. The transistors (the threshold voltages) chosen by this method are

- the pull-down and output transistors in the active write-driver AND gate,

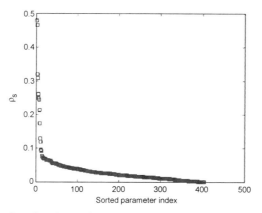

(a) Magnitudes of rank correlation between the statistical parameters and the write time of the SRAM column. Only a few parameters have a strong relationship with the write time

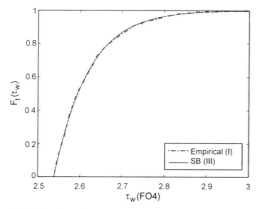

(b) GPD model of SRAM column write time from statistical blockade (6,314 simulations) compared with the empirical conditional CDF (100,000 simulations)

Figure 3.11. Results for the SRAM column test circuit

- the bitline pull-down transistors, and
- all transistors in the active 6T cell, except for M_{p2} (since node 2 is being pulled down in this case).

This selection coincides with a designer's intuition of the devices that would have the most impact on the write time.

y_σ is computed for increasing failure thresholds, using all three methods. We use $n_{MC} = 100{,}000$ simulated Monte Carlo points for methods I and II. For statistical blockade, method III, we filter these 100,000 points through the classifier in reduced dimensions, giving 5,314 candi-

τ_w (y_f) (FO4)	(I) Standard Monte Carlo	(II) GPD no filter	(III) SB (100K)	(III) SB (20K)	Gaussian approximation
2.7	2.966	2.986	3.010	2.990	3.364
2.8	3.367	3.373	3.390	3.425	3.898
2.9	3.808	3.743	3.747	3.900	4.432
3.0	∞	4.101	4.088	4.448	4.966
3.1	∞	4.452	4.416	5.138	5.499
3.2	∞	4.799	4.736	6.180	6.033
3.3	∞	5.147	5.049	–	6.567
3.4	∞	5.496	5.357	–	7.100
Number of simulations	100,000	100,000	6,314	2,046	20,000

Table 3.4. Prediction of failure probability as y_σ by methods I, II, III and by Gaussian approximation, for the SRAM column. The number of simulations for statistical blockade (SB) includes the 1,000 training samples. The write time values are shown in "fanout of 4" units

date tail point. On simulation, we finally obtain 1,077 true tail points. Table 3.4 compares the predictions by these three methods. We can see the close match between the accurate method II and statistical blockade, even though the total number of simulations is reduced from 100,000 to 6,314. The empirical method, once again, falls short of our needs, running out of data beyond $\tau_w = 2.9$ FO4. Figure 3.11(b) graphically shows the agreement between the conditional tail models extracted empirically and using statistical blockade.

We further reduce the Monte Carlo sample size for statistical blockade, to see if the simulation cost can be further reduced while maintaining accuracy. We use statistical blockade on only 20,000 Monte Carlo points, giving 1,046 filtered candidate tail points and 218 true tail points. However, the predictions (column 5) show large errors compared to our reference, method II. This suggests that a tail sample of only 218 is not sufficient to obtain a reliable model. We also use a Gaussian fit to 20,000 simulated Monte Carlo points for estimating y_σ. It is clear from the table that, in this case, a Gaussian fit under-estimates the failure probability, with the error increasing as we move to rarer events.

Comparing the statistics for the SRAM column in Table 3.4 with the statistics for the SRAM cell in Table 3.3, we can see that the distribution of write time has a larger spread for the SRAM column than for the SRAM cell. For example, the 4.8σ point for the SRAM cell is 2.7 FO4, while for the SRAM column, it is 3.2 FO4. The reason for this increased spread is that the variations in the leakage current of the entire column

Statistical Blockade 153

contribute significantly to the variation of the performance of any single cell. This shows that, in general, simulating variations in a single circuit, without modeling variations in its environment circuitry can lead to large errors in the estimated statistics.

3.3.4.3 Master–Slave Flip-Flop with Scan Chain

This last test case is a master-slave flip-flop with the scan chain component (MSFF). It is implemented in 45 nm technology, and is exactly the same as the one used in Sect. 1.7.1 for validating the SiLVR modeling method proposed in this thesis. Flip-flops are ubiquitous in digital circuits, and can be highly replicated in large chips. Here too, we are measuring the clock-output delay τ_{cq}. The flip-flop has a peculiarity in its rare event behavior. For large deviations in the statistical parameters, the flip-flop reaches metastable behavior and, using standard circuit simulators, we fail to see the flip-flop output converge to a stable low/high value. This leads to an undefined clock-output delay for some Monte Carlo points. We reject any such points without replacement in this experiment. Although these rejected points are also rare events, they distort the smoothness of the tail that is required to apply the EVT limit theorem, if not rejected. This still allows us to test the speed and tail modeling efficiency of statistical blockade, since we use the same rejection method across all estimation methods. In practice, the fraction of such undefined delay events can be estimated from the simulated points in statistical blockade and added to the failure probability estimated using the GPD model, to give the overall failure probability.

For methods I and II, we simulate $n_{MC} = 500{,}000$ Monte Carlo points. For statistical blockade (method III), we filter 100,000 Monte Carlo points to obtain 7,785 candidate tail points which, on simulating, yield 692 tail points. Note that here we have ignored any tail points for which the flip-flop output did not converge to a stable low/high value. Figure 3.12(b) compares the conditional tail CDFs from the empirical and statistical blockade models. We also compare predictions from these methods with those from a Gaussian fit to 20,000 simulated Monte Carlo points. Figure 3.12(a) shows a histogram of the delay values obtained from the 500,000-point Monte Carlo run. The extreme skewness and the heavy tail of the histogram suggest that a Gaussian fit would be grossly inaccurate.

Table 3.5 shows the estimates of y_σ computed by these four methods: we can clearly see the gross errors in the Gaussian estimates. In this case, we also see some discrepancy between the empirical and GPD-predicted values, that is larger than in the cases of the SRAM cell and column.

There can be two reasons for this:

154   FAST STATISTICAL ANALYSIS

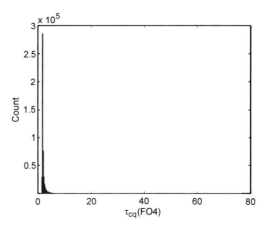

(a) Histogram of clock-output delay of the MSFF, showing a long heavy tail and high skewness

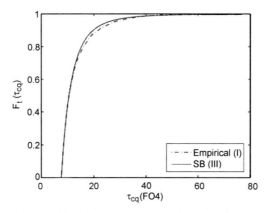

(b) GPD model of MSFF delay from statistical blockade (8,785 simulations) compared with the empirical conditional CDF (500,000 simulations)

Figure 3.12. Results for the MSFF test circuit

1) Due to the quite heavy tail, slight variations in the chosen tail samples can cause significant variations in the model.

2) The tail threshold of $t =$ the 99-th percentile might not be large enough to fit a GPD with near exactness; that is, the tail conditional CDF might not have converged to the GPD form.

It turns out that the actual reason is the second one. We further explore, and address, this problem in Sect. 3.4. The GPD fits do, however, capture the heavy tail of the distribution. To see this, compare Table 3.5 with the results for the SRAM cell in Table 3.3. A 20% increase in the SRAM write time, from 2.5 FO4 to 3 FO4, results in an increase of 2.424 in y_σ,

Statistical Blockade

τ_{cq} (y_f) (FO4)	(I) Standard Monte Carlo	(II) GPD *no* blockade filter	(III) Statistical blockade	Gaussian approximation
30	3.424	3.466	3.431	22.127
40	3.724	3.686	3.661	30.050
50	4.008	3.854	3.837	37.974
60	4.219	3.990	3.978	45.898
70	4.607	4.102	4.095	53.821
80	∞	4.199	4.195	61.745
90	∞	4.283	4.282	69.669
Number of simulations	500,000	500,000	8,785	20,000

Table 3.5. Prediction of failure probability as y_σ by methods I, II, III and by Gaussian approximation, for the MSFF. The number of simulations for statistical blockade includes the 1,000 training samples. The delay values are shown in "fanout of 4" units

while a similar percentage increase in the MSFF delay, from 50 FO4 to 60 FO4, increases y_σ by only 0.136, even though the increases are from similar probability levels (3.886σ for the SRAM cell, and 3.854σ for the MSFF).

In summary of these results, we see that statistical blockade provides an efficient way of sampling rare events and modeling their statistics. However, there are some issues that need to be addressed. First, the predictions may not be reliable for events that are very far out in the tail. Second, we saw some notable discrepancy between the empirical and GPD tail models for the case of the flip-flop: the exact reason for this is not yet obvious. The next section explores these issues in more detail and proposes enhancements to the statistical blockade method to address them.

3.4 Making Statistical Blockade Practical

Although statistical blockade provides us an effective method for sampling rare events and modeling their statistics, there are still some practical issues left unresolved by the algorithm in Sect. 3.3.3. We saw a glimpse of some of these issues in the results presented in Sect. 3.3.4. Let us look at these in more detail now.

3.4.1 Conditionals and Disjoint Tail Regions
3.4.1.1 The Problem
SRAM performance metrics are often computed for two states of the SRAM cell: while storing a 1, and while storing a 0. The final metric

value is then a maximum or a minimum of the vales for these two states. The presence of such *conditionals* (max, min) can result in *disjoint* tail regions in the statistical parameter space, making it difficult to use a single classifier to define the boundary of the tail region. Let us look at an example to illustrate this problem.

Consider the 6T SRAM cell. With technology scaling reaching nanometer feature sizes, sub-threshold and gate leakage become very significant. Particularly, for the large memory blocks seen today, the standby power consumption due to leakage can be intolerably high. Supply voltage (V_{dd}) scaling [KAdB02] is a powerful technique to reduce this leakage, whereby the supply voltage is reduced when the memory bank is not being accessed. However, lowering V_{dd} also makes the cell unstable, ultimately resulting in data loss at some threshold value of V_{dd}, known as the *data retention voltage* or DRV. Hence, the DRV of an SRAM cell is the lowest supply voltage that still preserves the data stored in the cell. DRV is computed as follows.

$$\text{DRV} = \max(\text{DRV}_0, \text{DRV}_1) \qquad (3.57)$$

where DRV_0 is the DRV when the cell is storing a 0, and DRV_1 it the DRV when it is storing a 1. If the cell is balanced (symmetric), with identical left and right halves, then $\text{DRV}_0 = \text{DRV}_1$. However, if there is any mismatch due to process variations, they become unequal. This creates the situation where the standard statistical blockade classification technique would fail because of the presence of disjoint tail regions.

Suppose we run a 1,000-point Monte Carlo, varying all the mismatch parameters in the SRAM cell according to their statistical distributions. This would give us distributions of values for DRV_0, DRV_1 and DRV. In certain parts of the mismatch parameter space $\text{DRV}_0 > \text{DRV}_1$, and in other parts, $\text{DRV}_0 < \text{DRV}_1$. This is clearly illustrated by Fig. 3.13(a): let us see how. Using the SiLVR method proposed in this thesis (Chap. 1), we extract the direction in the parameter space that has maximum impact on DRV_0. This direction is essentially the projection vector $\mathbf{w}_{1,\text{DRV}_0}$ for the first latent variable of DRV_0. The figure plots the simulated DRV_0 and DRV_1 values from the 1,000-point Monte Carlo run, along this direction; i.e., against the first latent variable d_{1,DRV_0}. It is clear that they are inversely related: one decreases as the other increases.

Now, let us take the maximum as in (3.57), and choose the classification threshold t_c equal to the 97-th percentile. Then we pick out the worst 3% points from the classifier training data and plot them against the same latent variable in Fig. 3.13(a), as red squares. Note that we have not trained the classifier yet, we are just looking at the points that the classifier would have to classify as being in the tail. We can clearly

Statistical Blockade

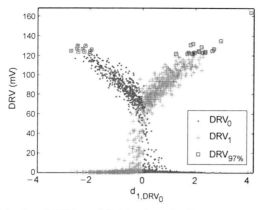

(a) Behavior of DRV_0 and DRV_1 along the direction of maximum variation in DRV_0. The worst 3% DRV values are plotted as *(red) squares*, clearly showing the disjoint tail regions (along this direction in the parameter space)

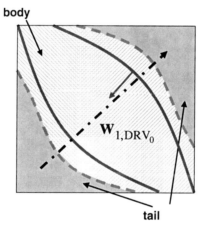

(b) A circuit metric (e.g., DRV) with two disjoint tail regions. The tail regions are separated from the body region by *dashed lines*. $\mathbf{w}_{1,DRV_0}$ is the direction of maximum variation of the circuit metric

Figure 3.13. Illustration of disjoint tail regions resulting from conditionals

see that these points (the red squares) lie in two disjoint parts of the parameter space. Since the true tail region defined by the tail threshold $t > t_c$ will be a subset of the classifier tail region (defined by t_c), it is obvious that the true tail region consists of two disjoint regions of the parameter space. This is illustrated with a two dimensional example in Fig. 3.13(b). The figure also shows the maximum impact direction vector, similar to the projection vector $\mathbf{w}_{1,DRV_0}$ extracted by SiLVR. Although this vector is different from $\mathbf{w}_{1,DRV_0}$ (which lies in a higher dimensional space) we mark it as $\mathbf{w}_{1,DRV_0}$ to make the relation obvious. The dark tail regions on the top-right and bottom-left corners of the parameter space correspond to the large DRV values shown as (red) squares in Fig. 3.13(a).

Such conditionals are very common for SRAM cell metrics, and hence, a classification strategy for such cases is essential for practical use of statistical blockade. We now propose such a strategy.

3.4.1.2 The Solution

Instead of building a single classifier for the tail of in (3.57), let us build two separate classifiers, one for the 97-th percentile (t_{c,DRV_0}) of DRV_0, and another for the 97-th percentile (t_{c,DRV_1}) of DRV_1. The generated Monte Carlo samples can then be filtered through both these classifiers: points classified as "body" by *both* the classifiers will be blocked, and the rest will be simulated. In the general case for arbitrary number of arguments in the conditional, let the circuit metric is given as

$$y = \max(y_0, y_1, \ldots). \quad (3.58)$$

The resulting general algorithm is then as follows:

1) Perform initial sampling to generate training data to build the classifiers, and estimate tail and classification thresholds, t_i and $t_{c,i}$, respectively, for each $y_i, i = 0, 1, \ldots$. Also estimate the tail threshold t for y.

2) For each argument, $y_i, i = 0, 1, \ldots$, of the conditional (3.58), build a classifier C_i at a classification threshold $t_{c,i}$ that is less than the corresponding tail threshold t_i.

3) Generate more points using Monte Carlo, but block the points classified as "body" by *all* the classifiers. Simulate the rest and compute y for the simulated points.

Hence, in the case of Fig. 3.13(b), we build a separate classifier for each of the two boundaries. The resulting classification boundaries are shown as solid lines. From the resulting simulated points, those with $y > t$ are chosen as tail points for further analysis; e.g., for computing a GPD model for the tail distribution of y. Note that this same algorithm can also be used for the case of multiple circuit metrics. Each metric would have its own thresholds and its own classifier, just like each argument in (3.58), the only difference being that we would not be computing any conditional.

3.4.2 Extremely Rare Events and Statistics
3.4.2.1 Extremely Rare Events

The GPD tail model can be used to make predictions regarding rare events that are farther out in the tail than any of the data we used to compute the GPD model. Indeed, this is the compelling reason for adopting the GPD model. However, as suggest by common intuition and the results presented in Sect. 3.3.4, we expect the statistical confidence in the estimates to decrease as we predict farther out in the tail. Equivalently, the variance of the predictions will probably increase as we move out in the tail. We can estimate this confidence or variance in two different ways:

1) *Empirically*: Suppose we run 50 runs of Monte Carlo with n_{MC} samples each and compute a GPD tail model from each run, using points that exceed some fixed threshold t. This gives us 50 slightly different pairs of the GPD parameters (ξ, β), one for each of 50 GPD models so computed. Then, we can compute variance and confidence intervals of any statistical metric using the 50 estimates obtained from these tail models.

2) *Using asymptotic variance*: Sect. 3.2.4 gives us expressions for the asymptotic covariance matrix $\Sigma_{\xi,\beta}$ of $(\hat{\xi}, \hat{\beta})$ estimated using probability-weighted moment matching: see (3.35). We can replace the exact values of (ξ, β) with the estimated $(\hat{\xi}, \hat{\beta})$ to obtain an approximate covariance matrix $\Sigma_{\hat{\xi},\hat{\beta}}$. For reasonable large number of tail samples used in the estimation, we can assume normal distribution of these estimates with mean $(\hat{\xi}, \hat{\beta})$ and covariance $\Sigma_{\hat{\xi},\hat{\beta}}$. Then, we can sample the GPD parameters from this distribution to compute different estimates of some statistical metric, which can be used to compute the variance. Here we need to build only one GPD model using a single Monte Carlo run of n_{MC} points. However, because of the assumption of the onset of asymptotic normal distribution and the approximation of the covariance matrix, we expect this method to show some error.

We use both these methods to compute 95% confidence intervals for the estimate of the $m\sigma$ point of the SRAM cell write time, where $m \in [3, 6]$. For the empirical method we use 50 Monte Carlo runs of $n_{MC} = 100{,}000$ points each, and compute GPD models with $t =$ the 99-percentile write time. This gives us 50 different estimates of the $m\sigma$ point. These estimates are shown in Fig. 3.14(a). As expected, the spread of the estimates increases as we extrapolate further with the GPD model. We then compute 95% confidence intervals of the $m\sigma$ point estimates

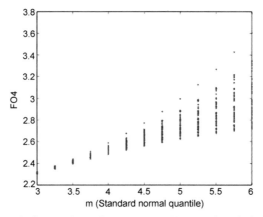

(a) The spread of $m\sigma$ point estimates across 50 runs of statistical blockade

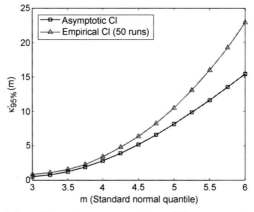

(b) 95% confidence intervals as a percentage of the mean value (empirical method) or the single estimate (asymptote method)

Figure 3.14. Variance in the estimates increases as we move further out in the tail

using these 50 models. Say we have 50 estimates $y_i(m), i = 1, \ldots, 50$ for the $m\sigma$ point. From these we can empirically compute the 97.5% percentile and 2.5% percentile points, $y_{97.5\%}(m)$ and $y_{2.5\%}(m)$, respectively. A 95% confidence interval $\kappa_{95\%}(m)$ can then be computed as

$$\kappa_{95\%}(m) = y_{97.5\%}(m) - y_{2.5\%}(m). \tag{3.59}$$

We express this confidence interval as a percentage of the mean of the estimates

$$\kappa'_{95\%}(m) = \frac{\kappa_{95\%}(m)}{\frac{1}{n}\sum_{i=1}^{50} y_i(m)}. \tag{3.60}$$

We also compute similar 95% confidence intervals using the second method, using 10,000 pairs of GPD parameter values sampled from the normal distribution with mean $(\hat{\xi}, \hat{\beta})$ and covariance $\Sigma_{\hat{\xi},\hat{\beta}}$. In this case we express the confidence interval as a percentage of the estimate $y(m)$ computed using the single GPD model $G_{\hat{\xi},\hat{\beta}}$. Figure 3.14(b) shows these percentage confidence intervals. Although there is some mismatch in the magnitudes of the two estimates of the 95% confidence interval, we see a common trend: the statistical confidence decreases as we move out in the tail. To keep the error within 5% with a confidence of 95% we should not be predicting farther than 4.28σ. For 10% error, we can go out to 4.95σ. Of course these numbers will change from circuit to circuit and performance metric to performance metric. The general inference is that we should not rely on the GPD tail model too far out from our data.

3.4.2.2 The Reason for Error in the MSFF Tail Model

Here we return to our MSFF test circuit from Sect. 3.3.4, where we saw some discrepancy between the empirical and GPD estimates for the failure probability expressed as y_σ. We will try to develop an explanation for this undesirable, although small, discrepancy. For this purpose we call on a common tool of graphical exploration of statistical data: the *sample mean excess plot*. [EKM03] reviews some properties of the mean excess plot. Here we focus on its properties in relation to the generalized Pareto distribution. The *mean excess function* for a given threshold y_f is defined as

$$e(y_f) = E(y - y_f | y > y_f); \qquad (3.61)$$

that is, the mean of exceedances over y_f. Plotting $e(y_f)$ against y_f gives us the mean excess plot. The *sample* mean excess function is the sample version of $e(y_f)$. For a given sample $\{y_i : i = 1, \ldots, n\}$, it is defined as

$$e_n(y_f) = \frac{\sum_{i=1}^{n}(y_i - y_f)_+}{|\{y_i : y_i > y_f\}|}, \quad \text{where } (\cdot)_+ = \max(\cdot, 0); \qquad (3.62)$$

that is, the sample mean of only the exceedances over y_f. A plot of $e_n(y_f)$ against y_f gives us the sample mean excess plot. The mean excess function of a GPD $G_{\xi,\beta}$ can be shown (see [EKM03]) to be a straight line given by

$$e(y_f) = \frac{\beta - \xi y_f}{1 + \xi}, \quad \text{for } y_f \in D(\xi, \beta), \qquad (3.63)$$

where $D(\xi, \beta)$ is as defined in Theorem 3.3. Hence, if the sample mean excess function of any data sample starts to follow roughly a straight line from some threshold, then it is an indication that the exceedances over

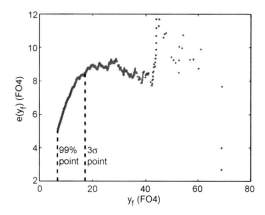

Figure 3.15. A sample mean excess plot for the MSFF circuit, showing the 99-th percentile and 3σ tail thresholds

that threshold follow a GPD. In fact, this feature of the mean excess plot can be employed to manually estimate an appropriate tail threshold.

Let us now look at the sample mean excess plot of the MSFF tail data ($\tau_{cq} \geq$ 99-percentile delay) from the 500,000-point Monte Carlo run. This is shown in Fig. 3.15. The plot suggests a good reason for the observed discrepancy in the estimated failure probabilities. It is clear from the plot that the tail defined by the $t = 99\%$ point has not converged close to a GPD form. Hence, the discrepancy could be a result of choosing a tail threshold that is not large enough. To test this, let us choose a threshold $t = 3\sigma$ point and fit the GPD model to exceedances over this t. Figure 3.15 suggests that this should show a better fit, since the sample mean excess function seems to be roughly a straight line from the 3σ threshold. The predictions of this new GPD model are shown in Table 3.6. We also reproduce columns 2 and 3 of Table 3.5 for comparison. As expected, we see more accurate predictions.

3.4.2.3 The Problem

For both the issues discussed above, the solution is to sample further out in the tail and use a higher tail threshold for building the GPD model of the tail. This is, of course, "easier said than done". Suppose we wish to support our GPD model with data up to the 6σ point. The failure probability of a 6σ value is roughly 1 part per *billion*, corresponding to a 99% chip yield requirement for a 10 Mb cache (with no error protection). This is definitely not an impractical requirement. However, for a 99% tail threshold, even a perfect classifier ($t_c = t$) will only reduce the number of simulations to an extremely large 10 million. If we decide to use a

Statistical Blockade

τ_{cq} (y_f) (FO4)	(I) Standard Monte Carlo	(II) GPD at 99-th percentile	(II) GPD at 3σ point
30	3.424	3.466	3.443
40	3.724	3.686	3.729
50	4.008	3.854	3.978
60	4.219	3.990	4.198
70	4.607	4.102	4.396
80	∞	4.199	4.574
90	∞	4.283	4.737

Table 3.6. Prediction of failure probability as y_σ using a GPD model (method II of Sect. 3.3.4) with the tail threshold t at the 99-th percentile and at the 3σ point

99.9999% threshold, the number of simulations will be reduced to a more practical 1,000 tail points (with a perfect classifier). However, we will need to simulate an extremely large number of points (≥ 1 million) to generate a classifier training set with at least one point in the tail region. In both cases, the circuit simulation counts are too high. We now describe a recursive formulation of statistical blockade that reduces this count drastically.

3.4.3 A Recursive Formulation of Statistical Blockade

Let us first assume that there are no conditionals. For a tail threshold equal to the a-th percentile, let us represent it as t^a, and the corresponding classification threshold as t_c^a. For this threshold, build a classifier C^a and generate sufficient points beyond the tail threshold, $y > t^a$, so that a *higher* percentile (t^b, t_c^b, $b > a$) can be estimated. For this new, higher threshold (t_c^b), a new classifier C^b is trained and a new set of tail points ($y > t^b$) is generated. This new classifier will block many more points than C^a, significantly reducing the number of simulations. This procedure is repeated to push the threshold out more until the tail region of interest is reached. The complete algorithm is shown in Algorithm 3.2.

The arguments to the algorithm are formulated a little differently from the basic statistical blockade algorithm (Algorithm 3.1). Instead of passing the tail and classification threshold probabilities (p_t, p_c), we pass a tail sample size n_t and a classification threshold probability function $p_c(p)$. The former is the number of tail points to be used finally to compute the GPD tail model. The latter is a function that returns the classification threshold probability for a given tail threshold probability. It is implicitly a function also of the classifier being used, since

Algorithm 3.2 The general recursive statistical blockade algorithm for efficient sampling of *extremely* rare events, in the presence of conditional induced disjoint tail regions

Require: initial sample size n_0 (e.g., 1,000); total sample size n; tail sample size n_t; function $p_c(p), p \in (0, 100)$; performance metric function $y = \max(y_0, y_1, \ldots)$
1: $\mathbf{X} = \text{MonteCarlo}(n_0)$
2: $n' = n_0$
3: $n_c = \max(n_t, 1000)$ // Classifier training set size at least 1,000
4: $\mathbf{Y} = \mathbf{f}_{\text{sim}}(\mathbf{X})$ // Simulate initial Monte Carlo sample
5: $\mathbf{y}_{\text{tail},i} = \mathbf{Y}_{\cdot,i}, i = 0, 1, \ldots$ // The i-th column contains values for y_i in $y = \max(y_0, y_1, \ldots)$
6: $\mathbf{X}_{\text{tail},i} = \mathbf{X}, i = 0, 1, \ldots$
7: **while** $n' < n$ **do**
8:     $\Delta n = \min(100 n', n) - n'$ // Number of points to filter in this recursion step
9:     $p_t = \frac{100 \Delta n}{n' + \Delta n}$; // Tail threshold is p_t-th percentile
10:     $n' = n' + \Delta n$ // Total number of points filtered at the end of this recursion step
11:     $\mathbf{X} = \text{MonteCarloNext}(\Delta n)$ // The next Δn points in the Monte Carlo sequence
12:     **for all** $i : y_i$ is an argument in $y = \max(y_0, y_1, \ldots)$ **do**
13:         $(\mathbf{X}_{\text{tail},i}, \mathbf{y}_{\text{tail},i}) = \text{GetWorst}(n_c, \mathbf{X}_{\text{tail},i}, \mathbf{y}_{\text{tail},i})$ // Get the n_t worst points
14:         $t = \text{Percentile}(\mathbf{y}_{\text{tail},i}, p_t)$
15:         $t_c = \text{Percentile}(\mathbf{y}_{\text{tail},i}, p_c(p_t))$
16:         $C_i = \text{BuildClassifier}(\mathbf{X}_{\text{tail},i}, \mathbf{y}_{\text{tail},i}, t_c)$
17:         $(\mathbf{X}_{\text{tail},i}, \mathbf{y}_{\text{tail},i}) = \text{GetGreaterThan}(t, \mathbf{X}_{\text{tail},i}, \mathbf{y}_{\text{tail},i})$ // Get the points with $y_i > t$
18:         $\mathbf{X}_{\text{cand},i} = \text{Filter}(C_i, \mathbf{X})$ // Candidate tail points for y_i
19:     **end for**
20:     $\mathbf{X} = [\mathbf{X}_{\text{cand},0}^T \; \mathbf{X}_{\text{cand},1}^T \ldots]^T$ // Union of *all* candidate tail points
21:     $\mathbf{Y} = \mathbf{f}_{\text{sim}}(\mathbf{X})$ // Simulate all candidate tail points
22:     $\mathbf{y}_{\text{cand},i} = \{\mathbf{Y}_{j,i} : \mathbf{X}_{j,\cdot} \in \mathbf{X}_{\text{cand},i}\}, i = 0, 1, \ldots$ // Extract the tail points for y_i
23:     $\mathbf{y}_{\text{tail},i} = [\mathbf{y}_{\text{tail},i}^T \; \mathbf{y}_{\text{cand},i}^T]^T$, $\mathbf{X}_{\text{tail},i} = [\mathbf{X}_{\text{tail},i}^T \; \mathbf{X}_{\text{cand},i}^T]^T$, $i = 0, 1, \ldots$ // All tail points till now
24: **end while**
25: $\mathbf{y}_{\text{tail}} = \text{MaxOverRows}([\mathbf{y}_{\text{tail},0} \; \mathbf{y}_{\text{tail},1} \ldots])$ // Compute the conditional
26: $\mathbf{y}_{\text{tail}} = \text{GetWorst}(n_t, \mathbf{y}_{\text{tail}})$
27: $(\xi, \beta) = \text{FitGPD}(\mathbf{y}_{\text{tail}} - \min(\mathbf{y}_{\text{tail}}))$

the error in the classifier will determine the appropriate safety margin. The functions that appear also in Algorithm 3.1 do the same work here, hence we do not reiterate their description. $\mathbf{f}_{\text{sim}}$ now returns multiple outputs: it computes the values of all the arguments of the conditional in $y = \max(y_0, y_1, \ldots)$. For example, in the case of DRV, it will return the values of DRV_0 and DRV_1. These values, for any one Monte Carlo point, are stored in one row of the result matrix $\mathbf{Y}$. The function $\texttt{MonteCarloNext}(\Delta n)$ returns the *next* Δn points in the sequence of

Algorithm 3.3 The recursive statistical blockade algorithm with fixed sequences for the tail and classification thresholds: $t = 99\%-, 99.99\%-, 99.9999\%-, \ldots$ points and $t_c = 97\%-, 99.97\%-, 99.9997\%, \ldots$ points. The total sample size is given by (3.64)

Require: initial sample size n_0 (e.g., 1,000); total sample size n; performance metric function $y = \max(y_0, y_1, \ldots)$
 1: $\mathbf{X} = \text{MonteCarlo}(n_0)$
 2: $n' = n_0$
 3: $\mathbf{Y} = \mathbf{f}_{\text{sim}}(\mathbf{X})$ // Simulate initial Monte Carlo sample
 4: $\mathbf{y}_{\text{tail},i} = \mathbf{Y}_{\cdot,i}, i = 0, 1, \ldots$ // The i-th column contains values for y_i in $y = \max(y_0, y_1, \ldots)$
 5: $\mathbf{X}_{\text{tail},i} = \mathbf{X}, i = 0, 1, \ldots$
 6: **while** $n' < n$ **do**
 7: $\quad \Delta n = 99n'$ // Number of points to filter in this recursion step
 8: $\quad n' = n' + \Delta n$ // Total number of points filtered at the end of this recursion step
 9: $\quad \mathbf{X} = \text{MonteCarloNext}(\Delta n)$ // The next Δn points in the Monte Carlo sequence
10: $\quad$ **for all** $i : y_i$ is an argument in $y = \max(y_0, y_1, \ldots)$ **do**
11: $\quad\quad (\mathbf{X}_{\text{tail},i}, \mathbf{y}_{\text{tail},i}) = \text{GetWorst}(1{,}000, \mathbf{X}_{\text{tail},i}, \mathbf{y}_{\text{tail},i})$ // Get the 1,000 worst points
12: $\quad\quad t = \text{Percentile}(\mathbf{y}_{\text{tail},i}, 99)$
13: $\quad\quad t_c = \text{Percentile}(\mathbf{y}_{\text{tail},i}, 97)$
14: $\quad\quad C_i = \text{BuildClassifier}(\mathbf{X}_{\text{tail},i}, \mathbf{y}_{\text{tail},i}, t_c)$
15: $\quad\quad (\mathbf{X}_{\text{tail},i}, \mathbf{y}_{\text{tail},i}) = \text{GetGreaterThan}(t, \mathbf{X}_{\text{tail},i}, \mathbf{y}_{\text{tail},i})$ // Get the points with $y_i > t$
16: $\quad\quad \mathbf{X}_{\text{cand},i} = \text{Filter}(C_i, \mathbf{X})$ // Candidate tail points for y_i
17: $\quad$ **end for**
18: $\quad \mathbf{X} = [\mathbf{X}_{\text{cand},0}^T \ \mathbf{X}_{\text{cand},1}^T \ldots]^T$ // Union of *all* candidate tail points
19: $\quad \mathbf{Y} = \mathbf{f}_{\text{sim}}(\mathbf{X})$ // Simulate all candidate tail points
20: $\quad \mathbf{y}_{\text{cand},i} = \{\mathbf{Y}_{j,i} : \mathbf{X}_{j,\cdot} \in \mathbf{X}_{\text{cand},i}\}, i = 0, 1, \ldots$ // Extract the tail points for y_i
21: $\quad \mathbf{y}_{\text{tail},i} = [\mathbf{y}_{\text{tail},i}^T \ \mathbf{y}_{\text{cand},i}^T]^T$, $\mathbf{X}_{\text{tail},i} = [\mathbf{X}_{\text{tail},i}^T \ \mathbf{X}_{\text{cand},i}^T]^T$, $i = 0, 1, \ldots$ // All tail points till now
22: **end while**
23: $\mathbf{y}_{\text{tail}} = \text{MaxOverRows}([\mathbf{y}_{\text{tail},0} \ \mathbf{y}_{\text{tail},1} \ldots])$ // Compute the conditional
24: $\mathbf{y}_{\text{tail}} = \text{GetWorst}(n_t, \mathbf{y}_{\text{tail}})$
25: $(\xi, \beta) = \text{FitGPD}(\mathbf{y}_{\text{tail}} - \min(\mathbf{y}_{\text{tail}}))$

---

points generated till now. The function GetWorst(n, **X**, **y**) returns the n worst values in the vector **y** and the corresponding rows of the matrix **X**. This functionality naturally extends to the two argument GetWorst(n, **y**). GetGreaterThan(t, **X**, **y**) returns the elements of **y** that are greater than t, along with the corresponding rows of **X**.

The function $p_c(p)$ is not easy to determine, hence we also present a less general version as Algorithm 3.3, which can be used immediately by any practitioner. Here, we restrict the total sample size n to be some

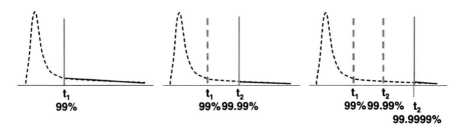

Figure 3.16. Recursive formulation of statistical blockade as in Algorithm 3.3

power of 100, times 1,000:

$$n = 100^j \cdot 1000, \quad j = 0, 1, \ldots \qquad (3.64)$$

Also, we fix $p_t = 99\%$ and $p_c = 97\%$. This will always give us 1,000 tail points to fit the GPD. The tail threshold t moves with every recursion step as

$t = 99\text{-th percentile}, 99.99\text{-th percentile}, 99.9999\text{-th percentile}, \ldots$

and the classification threshold as

$t_c = 97\text{-th percentile}, 99.97\text{-th percentile}, 99.9997\text{-th percentile}, \ldots$

The algorithms presented here are in iterative form, rather than recursive form. To see how the recursion works, suppose we want to estimate the 99.9999% tail. To generate points at and beyond this threshold, we first estimate the 99.99% point and use a classifier at the 99.97% point to generate these points efficiently. To build this classifier in turn, we first estimate the 99% point and use a classifier at the 97% point. Figure 3.16 illustrates this recursion on the PDF of any one argument in the conditional (3.58).

3.4.4 Experimental Results

We now test the recursive statistical blockade method on another SRAM cell test case, where we compute the data retention voltage (DRV) as in (3.57). In this case the SRAM cell is implemented in an industrial 90 nm process. Wang et al. [WSRC07] develop an analytical model for predicting the CDF of the DRV, that uses not more than 5,000 Monte Carlo points. The CDF is given as

$$F(y) = 1 - \text{erfc}(y_0) + \frac{1}{4}\text{erfc}^2(y_0), \quad \text{where } y_0 = \frac{\mu_0 + k(y - V_0)}{\sqrt{2}\sigma_0}, \qquad (3.65)$$

where y is the DRV value and erfc() is the complementary error function [PFTV92]. k is the sensitivity of the static noise margin (SNM) of the

SRAM cell to the supply voltage, computed using a DC sweep. μ_0 and σ_0 are the mean and standard deviation of the SNM (SNM$_0$), for a user-defined supply voltage V_0. SNM$_0$ is the SNM of the cell while storing a 0. These statistics are computed using a short Monte Carlo run of 1,500 to 5,000 sample points. We direct the reader to [WSRC07] for complete details regarding this analytical model of the DRV distribution. The q-th quantile can be estimated as

$$\mathrm{DRV}(q) = \frac{1}{k}(\sqrt{2}\sigma_0 \mathrm{erfc}^{-1}(2 - 2\sqrt{q}) - \mu_0) + V_0. \qquad (3.66)$$

Here DRV(q) is the supply voltage V$_{\mathrm{dd}}$ such that

$$P(\mathrm{DRV}(q) \leq \mathrm{V_{dd}}) = q. \qquad (3.67)$$

We compute the DRV quantiles as $m\sigma$ points, such that q is the cumulative probability for the value m from a standard normal distribution. We use five different methods to estimate the DRV quantiles for $m \in [3, 8]$:

1) *Analytical*: Use (3.66).

2) *Recursive statistical blockade without the GPD model*: Algorithm 3.3 is run for $n = 1$ billion. This results in three recursion stages, corresponding to total sample sizes of $n' = 100{,}000$, 10 million and 1 billion Monte Carlo points, respectively. The worst DRV value for these three recursion stages are estimates of the 4.26σ, 5.2σ and 6σ points, respectively.

3) *GPD model from recursive statistical blockade*: The 1,000 tail points from the last recursion stage of the recursive statistical blockade run are used to fit a GPD model, which is then used to predict the DRV quantiles.

4) *Normal*: A normal distribution is fit to data from a 1,000 point Monte Carlo run, and used to predict the DRV quantiles.

5) *Lognormal*: A lognormal distribution is fit to the same set of 1,000 Monte Carlo points, and used for the predictions.

The results are shown in Fig. 3.17. From the plots in the figure, we can immediately see that the recursive statistical blockade estimates are very close to the estimates from the analytical model. This shows the efficiency of the recursive formulation in reducing the error in predictions for events far out in the tail. Table 3.7 shows the number of circuit simulations performed at each recursion stage. The total number of simulations is 41,721. This is not small, but in comparison to standard Monte

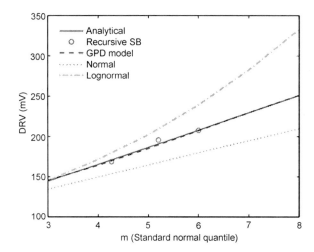

Figure 3.17. Estimates of DRV quantiles from five estimation methods. The GPD model closely fits the analytical model (3.65). The (*red*) *circles* show the worst DRV values from the three recursion stages of statistical blockade sampling. The normal and lognormal models are quite inaccurate

Recursion stage	Number of simulations
Initial	1,000
1	11,032
2	14,184
3	15,505
Total	41,721
Speedup over Monte Carlo	23,969×
Speedup over statistical blockade	719×

Table 3.7. Number of circuit simulation needed by recursive statistical blockade to generate a 6σ point

Carlo (1 billion simulations), and basic, non-recursive statistical blockade (approximately, 30 million with $t_c = 97$-th percentile) it is extremely fast. 41,721 simulations for DRV computation of a 6T SRAM cell can be completed in several hours on a single computer. With the advent of multi-core processors, the total simulation time can be drastically reduced with proper implementation.

Note that we can extend the prediction power to 8σ with the GPD model, without any additional simulations. Standard Monte Carlo would need over 1.5 quadrillion circuit simulations to generate a single 8σ point. For this case, the speedup over standard Monte Carlo is extremely large.

As expected, the normal and lognormal fits show large errors. The normal fit is unable to capture the skewness of the DRV distribution. On the other hand, the lognormal distribution has a heavier tail than the DRV distribution.

3.5 Future Work

Statistical blockade, in its recursive formulation, makes estimation of rare event statistics practical. This capability can be of immense use to designers of high capacity memories: SRAM, DRAM, non-volatile memories. Since it exploits rigorous limit theorems from extreme value theory, it has the unique capability of estimating the entire distribution of rare events, even with limited data. However, this is still an initial work, that brings up some questions that can be the focus of future research:

- What are the best tail and classification thresholds to use, so that the overall simulation cost is minimized? A solution to this question will probably depend on the characteristics of the specific tail distribution and the classifier being used. It may be possible to use the sample mean excess function to compute an appropriate threshold where roughly linear behavior, similar to a GPD, starts.

- Can fast Monte Carlo techniques like importance sampling and quasi-Monte Carlo be combined with statistical blockade for improvements in the quality of prediction? The extension to quasi-Monte Carlo seems obvious: just replace the pseudorandom generator with a low-discrepancy sequence generator. However, it is not clear if the existence of undesirable patterns in the QMC sequences would adversely affect the estimates of rare event statistics.

- How widely applicable is statistical blockade? Only adoption by circuit designers in the industry and widespread testing on industrial test cases will answer this question.

Chapter 4

Concluding Observations

Statistical analysis of circuits at many, if not all, stages of the design process is now inevitable. Recognizing this, this thesis proposed a set of novel algorithms that significantly improve over current capabilities for statistical analysis of custom circuits. In particular, first, we saw a new response surface modeling strategy called SiLVR that employs the concepts of projection pursuit and latent variable regression to create flexible, robust, yet compact models of the circuit response to variations. SiLVR has the attractive virtue of reducing the dimensionality to a few important variables, and capturing the designer's insight as quantitative measures of relative global sensitivity and input referred correlation. Second, a fast Monte Carlo sampling technique called quasi-Monte Carlo was proposed for statistical simulation of circuit performances. QMC uses deterministic low discrepancy sequences instead of the standard pseudorandom sequences of Monte Carlo. We showed how, using intelligent variable-dimension mapping, we can achieve significant speed or accuracy gains over the standard Monte Carlo technique. Third, an efficient method for sampling rare circuit events and reliably modeling their statistics was proposed. The method is called statistical blockade and solves a critical problem for memory designers: estimating the extremely low failure probabilities required for robust memory design.

Each chapter provided introductory discussion and concluding remarks for the algorithm proposed therein. To avoid unnecessary repetition, here we only discuss some common themes bridging the three core chapters. An overarching aspect of the work presented in this thesis is the use of true statistical techniques to solve the increasingly non-deterministic problems stemming from process variations in circuits. Linear model based statistical static timing analysis (SSTA)

[CS05][VRK+04a] is one successful example of methods that cast the statistical problem into a deterministic framework. Another example is the power optimization approach proposed in [MDO05]. However, such techniques tend to be restricted to very specific problems, and understandably so because they exploit some very specific problem characteristics to achieve the deterministic formulation. Domain-independent methods, however, are possible, and maintain as much generality as possible by not trying to analyze away the statistical nature of the problem. All the algorithms proposed in this thesis fall in this latter category. They accept the relevant problems as being statistical and use the methods of statistical inference to obtain the solutions they seek. The focus is to make this inference process as fast and accurate as possible. These algorithms show that a statistical, domain-independent attack on these difficult statistical problems of variations in circuits can lead to very fruitful results.

Because of the common strategy of using statistical inference while maintaining problem generality, the three proposed algorithms share some common technical aspects. All three methods try to extract some characteristics of the problem before applying the core solution techniques: for QMC, we use designer input or an initial Monte Carlo sample to extract information about important variables; for SiLVR, we use an initial Monte Carlo sample to fit the model to; and for statistical blockade, we use an initial Monte Carlo sample to compute estimates for the tail threshold and to train the classifier. This seems natural, since a general technique, like Monte Carlo, can be improved upon often by exploiting some more information about the problem. Of course, the extreme cases of such an approach are problem-specific techniques like linear model based SSTA, which are completely tailored for a particular, constrained set of problems. In this thesis we have started to work toward methods on the other end of this spectrum; that is, closer to completely general techniques.

It is inevitable that some generality will be lost if we try to exploit any problem characteristic. The key is to use techniques that can adapt to a large enough variety of problem characteristics, such that we can still handle most practical scenarios. This is the guiding principle in developing all the algorithms proposed here. The QMC flow exploits small effective dimension, thus restricting its domain of dominance over standard Monte Carlo to problems with reasonably small effective dimension. However, in practice, most circuit performances depend primarily on a small set of important variables. SiLVR again exploits a similar feature of the problem: a few latent variables will be enough to explain most of the circuit behavior. For the same reasons as in the case of QMC, this is not very restrictive in practice. Further, it has the tools to automatically extract all such required information about the problem. Statistical blockade targets high replication circuits, which by requirement tend to be small, simple circuits. Consequently, such circuits are likely to show some "well behaved" performance metrics allowing us to use simple, classical classifiers with some confidence. There are some common cases where which good behavior is lost: when there is a conditional, like max() in the computation of the performance metric. However, by handling each argument of the such conditional operators independently, we can reclaim this good behavior. Again, we see that although some generality is lost by using simple classifiers, the lost generality is not required often in practice. Hence, these algorithms solve somewhat different problems, but share much in their strategies of attacking these problems.

Many ideas proposed in this thesis derive inspiration from parallel problems existing in other technical fields. Nonlinear regression and feature selection in statistics and machine learning, leading to projection pursuit, and regression in very high dimensions with insufficient data in chemometrics leading to latent variable regression: these techniques form the intellectual seeds for SiLVR. Similarly, a parallel is explicitly drawn in Chap. 2 between circuit yield estimation and Asian option pricing, suggesting that techniques for fast statistical quadrature can be ported over from computational finance to circuit analysis. QMC is one popular quadrature used in computational finance, and also in other domains like statistical physics. These observations inspired the application of QMC to circuit problems. Of course, this application required its own tricks for ensuring effective use of QMC. Lastly, there are strong parallels between the problem of estimating extremely low failure probabilities of SRAM cells and the problem of estimating the statistics of catastrophic insurance claims in insurance risk analysis, or of disastrous floods in hydrology. The elegant theory of extreme values provides a sound footing for developing statistical models of all these rare events. Classification techniques from machine learning help make these models practical to build. Overall, this thesis demonstrates, with rigorous examples, that application of ideas from seemingly unrelated fields can lead to very new ways of attacking and solving problems in our field of interest.

Appendix A
Derivations of Variance Values for Test Functions in Sect. 2.6.1

Here we derive the exact values of the variance of f_c and the one dimensional variance of f_s, as defined in Sect. 2.6.1. The derivations for the one dimensional variance of f_c and the variance of f_s are not shown since they are very similar to these derivations.

A.1 Variance of f_c

We can write the integrand as

$$f_c(\mathbf{x}) = \left(\sum_{i=1}^{5} x_i\right)^2 = \sum_{i=1}^{5} x_i^2 + 2\sum_{i=1}^{4}\sum_{j=i+1}^{5} x_i x_j, \tag{A.1}$$

where $C^5 = [0,1]^5$. The integral of f_c is given by

$$\begin{aligned}
Q(f_c) &= \int_{C^5} f_c(\mathbf{x}) d\mathbf{x} \\
&= \int_{C^5} \left(\sum_{i=1}^{5} x_i^2 + 2\sum_{i=1}^{4}\sum_{j=i+1}^{5} x_i x_j\right) d\mathbf{x} \\
&= \sum_{i=1}^{5} \int_0^1 \cdots \int_0^1 x_i^2 dx_1 dx_2 dx_3 dx_4 dx_5 \\
&\quad + 2\sum_{i=1}^{4}\sum_{j=i+1}^{5} \int_0^1 \cdots \int_0^1 x_i x_j dx_1 dx_2 dx_3 dx_4 dx_5 \\
&= \sum_{i=1}^{5} \frac{x_i^3}{3} \prod_{j\neq i} x_j \bigg|_0^1 + 2\sum_{i=1}^{4}\sum_{j=i+1}^{5} \frac{x_i^2 x_j^2}{4} \prod_{k\neq i,j} x_k \bigg|_0^1 \\
&= 5\cdot\frac{1}{3} + 2\cdot\frac{5(5-1)}{2}\frac{1}{4} \\
&= \frac{20}{3}.
\end{aligned} \tag{A.2}$$

A. Singhee, R.A. Rutenbar, *Novel Algorithms for Fast Statistical Analysis of Scaled Circuits*, Lecture Notes in Electrical Engineering 46,
© Springer Science + Business Media B.V. 2009

The variance of f_c is then given by

$$\sigma^2(f_c) = \int_{C^5} (f_c(\mathbf{x}) - Q(f_c))^2 d\mathbf{x} = \int_{C^5} \left(\sum_{i=1}^{5} x_i^2 + 2\sum_{i=1}^{4}\sum_{j=i+1}^{5} x_i x_j - \frac{20}{3} \right)^2$$
$$= I_1 + I_2 + I_3, \tag{A.3}$$

where

$$I_1 = \int_{C^5} \left(\sum_{i=1}^{5} x_i^2 \right)^2 d\mathbf{x}, \tag{A.4}$$

$$I_2 = 4 \int_{C^5} \left(\sum_{i=1}^{5} x_i^2 \right) \left(\sum_{i=1}^{4}\sum_{j=i+1}^{5} x_i x_j - \frac{10}{3} \right) d\mathbf{x}, \tag{A.5}$$

$$I_3 = 4 \int_{C^5} \left(\sum_{i=1}^{4}\sum_{j=i+1}^{5} x_i x_j - \frac{10}{3} \right)^2 d\mathbf{x}. \tag{A.6}$$

Expanding I_1, we get

$$I_1 = \sum_{i=1}^{5} \int_{C^5} x_i^4 d\mathbf{x} + 2 \sum_{i=1}^{4}\sum_{j=i+1}^{5} \int_{C^5} x_i^2 x_j^2 d\mathbf{x}$$
$$= \sum_{i=1}^{5} \frac{x_i^5}{5} \prod_{j \neq i} x_j \bigg|_0^1 + 2 \sum_{i=1}^{4}\sum_{j=i+1}^{5} \frac{x_i^3 x_j^3}{9} \prod_{k \neq i,j} x_k \bigg|_0^1 \tag{A.7}$$
$$= 5 \cdot \frac{1}{5} + 2 \cdot 10 \cdot \frac{1}{9} \tag{A.8}$$
$$= \frac{29}{9}, \tag{A.9}$$

where the 10 in (A.8) is $= \frac{5(5-1)}{2}$, the number of distinct cross terms $x_i^2 x_j^2$, as in (A.1). Writing

$$T = \sum_{i=1}^{4}\sum_{j=i+1}^{5} x_i x_j, \tag{A.10}$$

which has 10 terms $x_i x_j$, and expanding I_2, we get

$$I_2 = 4 \int_{C^5} \left(T - \frac{10}{3} \right) \sum_{i=1}^{5} x_i^2 d\mathbf{x}$$
$$= 4 \left\{ \sum_{i=1}^{5} \int_{C^5} x_i^2 T d\mathbf{x} - \frac{10}{3} \sum_{i=1}^{5} \int_{C^5} x_i^2 d\mathbf{x} \right\}. \tag{A.11}$$

Plugging in the expansion for T, we get

$$\int_{C^5} x_i^2 T d\mathbf{x} = \int_{C^5} x_1^3 \sum_{i=2}^{5} x_i d\mathbf{x} + \int_{C^5} x_1^2 \sum_{i=2}^{4}\sum_{j=i+1}^{5} x_i x_j d\mathbf{x}$$
$$= \frac{x_1^4}{4} \bigg|_0^1 \sum_{i=2}^{5} \int_{C^4} x_i dx_2 dx_3 dx_4 dx_5$$
$$+ \frac{x_1^3}{3} \bigg|_0^1 \sum_{i=2}^{4}\sum_{j=i+1}^{5} \int_{C^4} x_i x_j dx_2 dx_3 dx_4 dx_5$$

Appendix A

$$= \frac{1}{4} \cdot 4 \cdot \frac{1}{2} + \frac{1}{3} \cdot 6 \cdot \frac{1}{4}$$
$$= 1. \qquad (A.12)$$

Also, the second term in (A.11),

$$\sum_{i=1}^{5} \int_{C^5} x_i^2 d\mathbf{x} = 5 \cdot \frac{1}{3} = \frac{5}{3}, \qquad (A.13)$$

as in (A.1). Substituting (A.12) and (A.13) into (A.11), we get

$$I_2 = 4\left\{ 5 \cdot 1 - \frac{10}{3} \cdot \frac{5}{3} \right\} = -\frac{20}{9}. \qquad (A.14)$$

Expanding I_3, and using T from (A.10), we get

$$I_3 = 4\left\{ \int_{C^5} TT d\mathbf{x} - \frac{20}{3} \int_{C^5} T d\mathbf{x} + \frac{100}{9} \int_{C^5} d\mathbf{x} \right\}. \qquad (A.15)$$

Let us look closely at TT. It has a total of $10 \times 10 = 100$ product terms, and these terms can be only of three types t_1, t_2, t_3:

$$t_1 \sim x_i^2 x_j^2, \quad i \neq j,$$
$$t_2 \sim x_i^2 x_j x_k \sim x_i x_j^2 x_k, \quad i \neq j \neq k, \qquad (A.16)$$
$$t_3 \sim x_i x_j x_k x_l, \quad i \neq j \neq k \neq l. \qquad (A.17)$$

We know that there are 10 possibilities for t_1 given that T has 10 terms of type $x_i x_j$. For t_2, if we fix some i, j then there are 3 choices for k such that $k \in \{1, \ldots, 5\}$ and $k \neq i, k \neq j$. For each such choice of k, we can square either i or j to obtain $x_i^2 x_j x_k$ or $x_i x_j^2 x_k$, respectively. Hence, for each choice of $\{i, j\}$ we have 6 distinct possibilities for t_2. Since there are 10 possibilities for $\{i, j\}$ such that $i, j \in \{1, \ldots, 5\}$ and $i \neq j$, we get a total of $10 \times 6 = 60$ t_2 terms. That leaves us with $100 - 10 - 60 = 30$ terms of type t_3. Here we are counting each permutation as one separate term. Now, the integrals of all t_1 terms are equal because of symmetry of integrand and the range C^5. Similarly, for all t_2 and all t_3 terms. Since,

$$\int_{C^5} x_1^2 x_2^2 d\mathbf{x} = \left. \frac{x_1^3 x_2^3}{9} x_3 x_4 x_5 \right|_0^1 = \frac{1}{9}, \qquad (A.18)$$

$$\int_{C^5} x_1^2 x_2 x_3 d\mathbf{x} = \left. \frac{x_1^3 x_2^2 x_3^2}{12} x_4 x_5 \right|_0^1 = \frac{1}{12}, \qquad (A.19)$$

$$\int_{C^5} x_1 x_2 x_3 x_4 d\mathbf{x} = \left. \frac{x_1^2 x_2^2 x_3^2 x_4^2}{16} x_5 \right|_0^1 = \frac{1}{16}, \qquad (A.20)$$

we can write

$$\int_{C^5} TT d\mathbf{x} = 10 \int_{C^5} x_1^2 x_2^2 d\mathbf{x} + 60 \int_{C^5} x_1^2 x_2 x_3 d\mathbf{x} + 30 \int_{C^5} x_1 x_2 x_3 x_4 d\mathbf{x}$$
$$= 10 \cdot \frac{1}{9} + 60 \cdot \frac{1}{12} + 30 \cdot \frac{1}{16} = \frac{575}{72}. \qquad (A.21)$$

For the second term for I_3 in (A.15), we can write

$$\int_{C^5} T d\mathbf{x} = 10 \int_{C^5} x_i x_j d\mathbf{x} = 10 \cdot \frac{1}{4} = \frac{5}{2}, \qquad (A.22)$$

and for the third term,

$$\int_{C^5} d\mathbf{x} = 1. \qquad (A.23)$$

Plugging these, and (A.20) into (A.15), we then get

$$I_3 = 4\left\{\frac{575}{72} - \frac{20}{3} \cdot \frac{5}{2} + \frac{100}{9}\right\} = \frac{175}{18}. \qquad (A.24)$$

Hence, using (A.9), (A.14) and (A.24) in (A.3), we get

$$\sigma^2(f_c) = \frac{29}{9} - \frac{20}{9} + \frac{175}{18} = \frac{193}{18}. \qquad (A.25)$$

A.2 One Dimensional Variance of f_s

We write f_s as

$$f_s(\mathbf{x}) = f_c(\mathbf{x}) - f_a(\mathbf{x}) = 2\sum_{i=1}^{4}\sum_{j=i+1}^{5} x_i x_j. \qquad (A.26)$$

Define $f_{\{i\}}$ as the ANOVA component of f_s that is a function of only x_i. From (2.89), this is given as

$$f_{\{i\}}(\mathbf{x}) = \int_{C-\{i\}} (f_s(\mathbf{x}) - f_\emptyset(\mathbf{x})) d\mathbf{x}_{-\{i\}}, \qquad (A.27)$$

where

$$-\{i\} = \{1, \ldots, 5\} - \{i\} \qquad (A.28)$$

is the complementary set of $\{i\}$. $f_\emptyset(\mathbf{x})$ is the same as $Q(f_s)$, the integral of f_s, and we can write f_s as

$$f_s(\mathbf{x}) = 2x_i \sum_{j \in -\{i\}} x_j + 2 \sum_{k \in -\{i\}} \sum_{l \in -\{i\}: l > k} x_k x_l. \qquad (A.29)$$

Using this in (A.27), we can write

$$f_{\{i\}}(\mathbf{x}) = 2x_i \sum_{j \in -\{i\}} \int_{C-\{i\}} x_j d\mathbf{x}_{-\{i\}} + 2 \sum_{k \in -\{i\}} \sum_{l \in -\{i\}: l > k} \int_{C-\{i\}} x_k x_l d\mathbf{x}_{-\{i\}}$$

$$- Q(f_s) \int_{C-\{i\}} d\mathbf{x}_{-\{i\}}. \qquad (A.30)$$

By simple integration, as in the beginning of the previous section, we can show that $Q(f_s) = 5$. Integrating (A.29) and using this, we get

$$f_{\{i\}}(\mathbf{x}) = 2x_i \left(4 \cdot \frac{1}{2}\right) + 2 \cdot \frac{4(4-1)}{2} \frac{1}{4} - 5$$

$$= 4x_i - 2. \qquad (A.31)$$

The variance of $f_{\{i\}}$ is denoted by $\sigma^2_{\{i\}}$, and is given by

$$\sigma^2_{\{i\}} = \int_{C^5} (f_{\{i\}}(\mathbf{x}))^2 d\mathbf{x}. \qquad (A.32)$$

Appendix A

Using (A.31), we get

$$\begin{aligned}
\sigma_{\{i\}}^2 &= \int_0^1 (4x_i - 2)^2 dx_i \\
&= \int_0^1 (16x_i^2 - 8x_i + 4) dx_i \\
&= \frac{16}{3} - 8 + 4 \\
&= \frac{4}{3}.
\end{aligned} \qquad (A.33)$$

The total variance from one dimensional components, σ_1^2, is given as

$$\sigma_1^2 = \sum_{i=1}^{5} \sigma_{\{i\}}^2 = 5 \cdot \frac{4}{3} = \frac{20}{3}. \qquad (A.34)$$

References

[ABG98] P. Acworth, M. Broadie, and P. Glasserman. A comparison of some Monte Carlo and quasi-Monte Carlo techniques for option pricing. In H. Niederreiter, P. Hellekalek, G. Larcher, and P. Zinterhof, editors, *Monte Carlo and Quasi-Monte Carlo Methods 1996*, pages 1–18. Springer, New York, 1998.

[Ack] P. J. Acklam. An algorithm for computing the inverse normal cumulative distribution function. http://home.online.no/~pjacklam/notes/invnorm/.

[Ada75] R. A. Adams. *Sobolev Spaces*. Academic Press, New York, 1975.

[AGW94] K. J. Antreich, H. E. Graeb, and C. U. Weiser. Circuit analysis and optimization driven by worst-case distances. *IEEE Trans. Computer-Aided Design*, 13(1):57–71, 1994.

[AMH91] H. L. Abdel-Malik and A.-K. S. O. Hassan. The ellipsoidal technique for design centering and region approximation. *IEEE Trans. Computer-Aided Design*, 10(8):1006–1014, 1991.

[AS79] I. A. Antanov and V. M. Saleev. An economic method of computing LP_τ-sequences. *U.S.S.R. Comp. Math. and Math. Phys.*, 19:252–256, 1979 (English translation).

[Bak59] N. S. Bakhvalov. On approximate calculation of integrals. *Vestnik Moskow. Gos. Univ., Ser. Mat. Mekh. Astronom. Fiz. Khim.*, 4:3–18, 1959 (in Russian).

[Bar89] A. R. Barron. Statistical properties of artificial neural networks. In *Proc. 28th Conf. Decision and Control, December 1989*.

[Bar93] A. R. Barron. Universal approximation bounds for superpositions of a sigmoidal function. *IEEE Trans. Inform. Theory*, 39(3):930–945, 1993.

[BBT97] A. M. Bruckner, J. B. Bruckner, and B. S. Thompson. *Real Analysis*. Prentice–Hall, Englewood Cliffs, 1997.

[BdH74] A. A. Balkema and L. de Haan. Residual life time at great age. *Ann. Prob.*, 2(5):792–804, 1974.

[BF88] P. Bratley and B. L. Fox. Algorithm 659: implementing Sobol's quasirandom sequence generator. *ACM Trans. Math. Soft.*, 14(1):88–100, 1988.

[BFN92] P. Bratley, B. L. Fox, and H. Niederreiter. Implementation and tests of low-discrepancy sequences. *ACM Trans. Modeling Comp. Sim.*, 2(3):195–213, 1992.

[BM58] G. E. P. Box and M. E. Muller. A note on the generation of random normal deviates. *Ann. Math. Stats.*, 29:610–611, 1958.

[BMM99] G. Baffi, E. B. Martin, and A. J. Morris. Non-linear projection to latent structures revisited (the neural network PLS algorithm). *Comp. Chem. Engg.*, 23(9):1293–1307, 1999.

[BN01] M. Burger and A. Neubauer. Error bounds for approximation with neural networks. *J. Approx. Theory*, 112:235–250, 2001.

[BS06] A.-L. Boulesteix and K. Strimmer. Partial least squares: a versatile tool for the analysis of high-dimensional genomic data. *Brief. Bioinform.*, 8(1):32–44, 2006.

[BSUM99] H. Banba, H. Shiga, A. Umezawa, and T. Miyaba. A CMOS bandgap reference circuit with sub-1-v operation. *IEEE J. Solid-State Circuits*, 34(5):670–674, 1999.

[BTM01] A. J. Bhavnagarwala, X. Tang, and J. D. Meindl. The impact of intrinsic device fluctuations on CMOS SRAM cell stability. *IEEE J. Solid-State Circuits*, 36(4):658–665, 2001.

[Bur98] C. J. C. Burges. A tutorial on support vector machines for pattern recognition. *Data Min. Knowl. Discov.*, 2(2):121–167, 1998.

[BVM96] A. J. Burnham, R. Viveros, and J. F. MacGregor. Frameworks for latent variable multivariate regression. *J. Chemometrics*, 20:31–45, 1996.

[CC05] B. H. Calhoun and A. Chandrakasan. Analyzing static noise margin for sub-threshold SRAM in 65 nm CMOS. In *Proc. Europ. Solid State Cir. Conf.*, 2005.

[CL92] C. K. Chui and X. Li. Approximation by ridge functions and neural networks with one hidden layer. *J. Approx. Theory*, 70:131–141, 1992.

[CLM96] C. K. Chui, X. Li, and H. N. Mhaskar. Limitations of the approximation capabilities of neural networks with one hidden layer. *Adv. Comp. Math.*, 5:233–243, 1996.

[CLR01] T. H. Cormen, C. E. Leiserson, and R. L. Rivest. *Introduction to Algorithms*, 2nd edition. MIT Press, Cambridge, 2001.

[CMO97] R. E. Caflisch, W. Morokoff, and A. Owen. Valuation of mortgage backed securities using Brownian bridges to reduce effective dimension. *J. Comp. Finance*, 1(1):27–46, 1997.

[Coo99] R. Cools. Monomial cubature rules since "Stroud": a compilation – part 2. *J. Comput. Appl. Math.*, 112:21–27, 1999.

[CS96] F. M. Coetzee and V. L. Stonick. On the uniqueness of weights in single-layer perceptron. *IEEE Trans. Neural Networks*, 7(2):318–325, 1996.

[CS05] H. Chang and S. Sapatnekar. Statistical timing under spatial correlations. *IEEE Trans. Computer-Aided Design*, 24(9):1467–1482, 2005.

[Cyb89] G. Cybenko. Approximation by superpositions of sigmoidal functions. *Math. Control Signals Systems*, 2:303–314, 1989.

[CYMSC85] P. Cox, P. Yang, S. S. Mahant-Shetti, and P. Chatterjee. Statistical modeling for efficient parametric yield estimation of MOS VLSI circuits. *IEEE Trans. Electron Devices*, 32(2):471–478, 1985.

[DFK93]   S. W. Director, P. Feldmann, and K. Krishna. Statistical integrated circuit design. *IEEE J. Solid-State Circuits*, 28(3):193–202, 1993.

[dH90]   L. de Haan. Fighting the arch-enemy with mathematics. *Statist. Neerlandica*, 44:45–68, 1990.

[DJRS85] D. Donoho, I. Johnstone, P. Rousseeuw, and W. Stahel. Projection pursuit (discussion). *Ann. Stats.*, 13(2):496–500, 1985.

[DS84]   P. Diaconis and M. Shahshahani. On nonlinear functions of linear combinations. *SIAM J. Sci. Statist. Comput.*, 5(1):175–191, 1984.

[DS96]   J. E. Dennis, Jr. and R. B. Schnabel. *Numerical Methods for Unconstrained Optimization and Nonlinear Equations.* SIAM, Philadelphia, 1996.

[DT82]   P. Davies and M. K.-S. Tso. Procedures for reduced-rank regression. *Appl. Stats.*, 31(3):244–255, 1982.

[EIH02]   T. Ezaki, T. Izekawa, and M. Hane. Investigation of random dopant fluctuation induced device characteristics variation for sub-100 nm CMOS by using atomistic 3d process/device simulator. In *Proc. IEEE Int. Electron Devices Meeting, 2002*.

[EKM97]   P. Embrechts, C. Klüppelberg, and T. Mikosch. *Modelling Extremal Events.* Springer, Berlin, 1997.

[EKM03]   P. Embrechts, C. Klüppelberg, and T. Mikosch. *Modelling Extremal Events for Insurance and Finance*, 4th printing edition. Springer, Berlin, 2003.

[Eli94]   N. J. Elias. Acceptance sampling: an efficient, accurate method for estimating and optimizing parametric yield. *IEEE J. Solid-State Circuits*, 29(3):323–327, 1994.

[Fau82]   H. Faure. Discrépance de suites associées à un système de numération (en dimension s). *Acta Arith.*, 41:337–351, 1982 (in French).

[FD93]   P. Feldmann and S. W. Director. Integrated circuit quality optimization using surface integrals. *IEEE Trans. Computer-Aided Design*, 12(12):1868–1879, 1993.

[FH97]   F. D. Foresee and M. T. Hagan. Gauss–Newton approximation to Bayesian learning. In *Proc. Int. Conf. Neural Networks, June 1997*.

[Fis06]   G. S. Fishman. *A First Course in Monte Carlo*. Duxbury, N. Scituate, 2006.

[FL06]   Z. Feng and P. Li. Performance-oriented statistical parameter reduction of parameterized systems via reduced rank regression. In *Proc. IEEE/ACM Int. Conf. on CAD, November 2006*.

[Fox86]   B. L. Fox. Algorithm 647: implementation and relative efficiency of quasirandom sequence generators. *ACM Trans. Math. Soft.*, 12(4):362–376, 1986.

[Fox99]   B. L. Fox. *Strategies for Quasi-Monte Carlo*. Kluwer Academic, New York, 1999.

[Fri84]   J. H. Friedman. A variable span smoother. Dept. of Statistics Tech. Report LCS 05, Stanford Univ., 1984.

[FS81]   J. H. Friedman and W. Stuetzle. Projection pursuit regression. *J. Amer. Stat. Assoc.*, 76(376):817–823, 1981.

[FT28] R. A. Fisher and L. H. C. Tippett. Limiting forms of the frequency distribution of the largest or smallest member of a sample. *Proc. Cambridge Philos. Soc.*, 24:180–190, 1928.

[FT02] H. Faure and S. Tezuka. Another random scrambling of digital (t,s)-sequences. In K.-T. Fang, F. J. Hickernell, and H. Niederreiter, editors, *Monte Carlo and Quasi-Monte Carlo Methods 2000*, pages 242–256. Springer, New York, 2002.

[FTIW99] D. J. Frank, Y. Taur, M. Ieong, and H.-S. P. Wong. Monte Carlo modeling of threshold variation due to dopant fluctuation. In *Proc. Int. Symp. VLSI Tech.*, 1999.

[Fun89] K. Funahashi. On the approximate realization of continuous mappings by neural networks. *Neural Networks*, 2:183–192, 1989.

[FW94] K.-T. Fang and Y. Wang. *Number Theoretic Methods in Statistics*. Chapman and Hall, London, 1994.

[GG98] T. Gerstner and M. Griebel. Numerical integration using sparse grids. *Numerical Algorithms*, 18(3–4):209–232, 1998.

[GH04] P. Gupta and F.-L. Heng. Toward a systematic-variation aware timing methodology. In *Proc. IEEE/ACM Design Autom. Conf.*, June 2004.

[GJLM01] P. R. Gray, P. J. Jurst, S. H. Lewis, and R. G. Meyer. *Analysis and Design of Analog Integrated Circuits*, 4th edition. Wiley, New York, 2001.

[GJP95] F. Girosi, M. Jones, and T. Poggio. Regularization theory and neural network architectures. *Neural Computation*, 7(2):219–269, 1995.

[GL96] G. Golub and C. Loan. *Matrix Computations*. JHU Press, Baltimore, 1996.

[Gla04] P. Glasserman. *Monte Carlo Methods in Financial Engineering*. Springer, Berlin, 2004.

[Gne43] B. Gnedenko. Sur la distribution limite du terme maximum d'une aleatoire. *Ann. Math.*, 44(3):423–453, 1943.

[Gri93] S. D. Grimshaw. Computing maximum likelihood estimates for the generalized Pareto distribution. *Technometrics*, 35(2):185–191, 1993.

[Hal60a] J. H. Halton. On the efficiency of certain quasi-random sequences of points in evaluating multi-dimensional integrals. *Numerische Mathematik*, 2:84–90, 1960.

[Hal60b] J. H. Halton. On the efficiency of certain quasi-random sequences of points in evaluating multi-dimensional integrals. *Numerische Mathematik*, 2:84–90, 1960.

[Hal89] P. Hall. On projection pursuit regression. *Ann. Stats.*, 17(2):573–588, 1989.

[Ham60] J. M. Hammersley. Monte Carlo methods for solving multivariate problems. *Ann. New York Acad. Sci.*, 86:844–874, 1960.

[HC71] R. V. Hogg and A. T. Craig. *Introduction to Mathematical Statistics*, 3rd edition. MacMillan, London, 1971.

[Hei94] S. Heinrich. Random approximation in numerical analysis. In K. D. Bierstedt, A. Pietsch, W. M. Ruess, and D. Vogt, editors, *Functional Analysis*, Marcel Dekker, New York, pages 123–171, 1994.

[Hei96] S. Heinrich. Complexity theory of Monte Carlo algorithms. *Lec. Appl. Math.*, 32:405–419, 1996.

References

[Hes03] T. C. Hesterberg. Advances in importance sampling. Dept. of Statistics, Stanford University, 1988, 2003.

[HH03] H. S. Hong and F. J. Hickernell. Algorithm 823: implementing scrambled digital sequences. *ACM Trans. Math. Soft.*, 29(2):95–109, 2003.

[HHLL00] F. J. Hickernell, H. S. Hong, P. L'Ecuyer, and C. Lemieux. Extensible lattice sequences for quasi-Monte Carlo quadrature. *SIAM J. Sci. Comp.*, 22(3):1117–1138, 2000.

[Hic98] F. J. Hickernell. A generalized discrepancy and quadrature error bound. *Math. Comp.*, 67(221):299–322, 1998.

[HIE03] M. Hane, T. Ikezawa, and T. Ezaki. Atomistic 3d process/device simulation considering gate line-edge roughness and poly-si random crystal orientation effects. In *Proc. IEEE Int. Electron Devices Meeting, 2003*.

[Hla61] E. Hlawka. Functionen von beschränkter variation in der theori der gleichverteilung. *Ann. Mat. Pura Appl.*, 54:325–333, 1961 (in German).

[HLT83] D. E. Hocevar, M. R. Lightner, and T. N. Trick. A study of variance reduction techniques for estimating circuit yields. *IEEE Trans. Computer-Aided Design*, 2(3):279–287, 1983.

[HM94] M. T. Hagan and M. B. Menhaj. Training feedforward networks with the Marquardt algorithm. *IEEE Trans. Neural Networks*, 5(6):989–993, 1994.

[Hos86] J. R. M. Hosking. The theory of probability weighted moments. IBM Research Report, RC12210, 1986.

[HSW89] K. Hornik, M. Stinchcombe, and H. White. Multilayer feedforward networks are universal approximators. *Neural Networks*, 2:359–366, 1989.

[HTF01] T. Hastie, R. Tibshirani, and J. Friedman. *The Elements of Statistical Learning: Data Mining, Inference, and Prediction*. Springer, Berlin, 2001.

[Hub85] P. J. Huber. Projection pursuit. *Ann. Stats.*, 13(2):435–475, 1985.

[HW81] L. K. Hua and Y. Wang. *Applications of Number Theory to Numerical Analysis*. Springer, Berlin, 1981.

[HW87] J. R. M. Hosking and J. R. Wallis. Parameter and quantile estimation for the generalized Pareto distribution. *Technometrics*, 29(3):339–349, 1987.

[IM88] B. Irie and S. Miyake. Capabilities of three-layered perceptrons. In *Int. Conf. Neural Networks, 1988*.

[Ism93] C. Michael, M. I. Ismael. *Statistical Modeling for Computer-Aided Design of Mos VLSI Circuits*. Springer, Berlin, 1993.

[JK03] S. Joe and F. Y. Kuo. Remark on algorithm 659: implementing Sobol's quasirandom sequence generator. *ACM Trans. Math. Soft.*, 29(1):49–57, 2003.

[Joa99] T. Joachims. Making large-scale SVM learning practical. In B. Schölkopf, C. Burges, and A. Smola, editors, *Advances in Kernel Methods – Support Vector Learning*. MIT Press, Cambridge, 1999.

[Joh55] F. John. *Plane Waves and Spherical Means Applied to Partial Differential Equations*. Interscience Publishers, New York, 1955.

[Jon87] L. K. Jones. On a conjecture of Huber concerning the convergence of projection pursuit regression. *Ann. Stats.*, 15(2):880–882, 1987.

[KAdB02]   R. K. Krishnamurthy, A. Alvandpour, V. De, and S. Borkar. High-performance and low-power challenges for sub-70 nm microprocessor circuits. In *Proc. Custom Integ. Circ. Conf., 2002*.

[Kar33]   J. Karamata. Sur un mode de croissance régulière. Théorèmes fondamentaux. *Bull. Soc. Math. France*, 61:55–62, 1933.

[Kie61]   J. Kiefer. On large deviations of the empirical d. f. of vector chance variables and a law of the iterated logarithm. *Pacific J. Math.*, 11:649–660, 1961.

[KJN06]   R. Kanj, R. Joshi, and S. Nassif. Mixture importance sampling and its application to the analysis of SRAM designs in the presence of rare event failures. In *Proc. IEEE/ACM Design Autom. Conf., 2006*.

[KN74]   L. Kuipers and H. Niederreiter. *Uniform Distribution of Sequences*. Wiley, New York, 1974.

[Kun95]   K. Kundert. *The Designer's Guide to SPICE and Spectre®*. Springer, Berlin, 1995.

[LGXP04]   X. Li, P. Gopalakrishnan, Y. Xu, and L. T. Pileggi. Robust analog/RF circuit design with projection-based posynomial modeling. In *Proc. IEEE/ACM Int. Conf. on CAD, 2004*.

[Lig92]   W. Light. Ridge functions, sigmoidal functions and neural networks. In E. W. Cheney, C. K. Chui, and L. L. Schumaker, editors, *Approximation Theory, VII*. Academic Press, San Diego, 1992

[LJC+88]   W. Liu, X. Jin, J. Chen, M.-C. Jeng, Z. Liu, Y. Cheng, K. Chen, M. Chan, K. Hui, J. Huang, R. Tu, P. Ko, and C. Hu. Bsim 3v3.2 mosfet model users' manual. Univ. California, Berkeley, Tech. Report No. UCB/ERL M98/51, 1988.

[LL02]   P. L'Ecuyer and C. Lemieux. A survey of randomized quasi-Monte Carlo methods. In M. Dror, P. L'Ecuyer, and F. Szidarovski, editors, *Modeling Uncertainty: An Examination of Stochastic Theory, Methods, and Applications*, pages 419–474. Kluwer Academic, New York, 2002.

[LLP04]   J. Le, X. Li, and L. T. Pileggi. STAC: statistical timing analysis with correlation. In *Proc. IEEE/ACM Design Autom. Conf., June 2004*.

[LLPS05]   X. Li, J. Le, L. T. Pileggi, and A. Stojwas. Projection-based performance modeling for inter/intra-die variations. In *Proc. IEEE/ACM Int. Conf. on CAD, November 2005*.

[Lo77]   M. Loéve. *Probability Theory I & II*, 4th edition. Springer, Berlin, 1977.

[LP93]   V. Ya. Lin and A. Pinkus. Fundamentality of ridge functions. *J. Approx. Theory*, 75:295–311, 1993.

[LS75]   B. F. Logan and L. A. Shepp. Optimal reconstruction of a function from its projections. *Duke Math. J.*, 42:645–659, 1975.

[Mac92]   D. J. C. MacKay. A practical Bayesian framework for backpropagation networks. *Neural Computation*, 4(3):448–472, 1992.

[Mai99]   V. E. Maiorov. On best approximation by ridge functions. *J. Approx. Theory*, 99:68–94, 1999.

[Mar63]   D. Marquardt. An algorithm for least squares estimation of non-linear parameters. *J. Soc. Indust. Appl. Math.*, 11:431–441, 1963.

[Mat98]   J. Matoušek. On the l_2-discrepancy for anchored boxes. *J. Complexity*, 14(4):527–556, 1998.

References

[MBC79] M. D. McKay, R. J. Beckman, and W. J. Conover. A comparison of three methods for selecting values of input variables in the analysis of output from a computer code. *Technometrics*, 21(2):239–245, 1979.

[MBJ99] K. Morik, P. Brockhausen, and T. Joachims. Combining statistical learning with a knowledge-based approach – a case study in intensive care monitoring. In *Proc. 16th Int. Conf. Machine Learning, 1999*.

[MC94] W. J. Morokoff and R. E. Caflisch. Quasi-random sequences and their discrepancies. *SIAM J. Sci. Comp.*, 15(6):1251–1279, 1994.

[MC95] W. J. Morokoff and R. E. Caflisch. Quasi-Monte Carlo integration. *J. Comput. Phys.*, 122(2):218–230, 1995.

[MC96] B. Moskowitz and R. E. Caflisch. Smoothness and dimension reduction in quasi-Monte Carlo methods. *Math. Comput. Modelling*, 23(8/9):37–54, 1996.

[MDO05] M. Mani, A. Devgan, and M. Orshansky. An efficient algorithm for statistical minimization of total power under timing yield constraints. In *Proc. IEEE/ACM Design Autom. Conf., 2005*.

[Mer73] R. C. Merton. Theory of rational option pricing. *The Bell J. Econ. Management Science*, 4(1):141–183, 1973.

[Mha92] H. N. Mhaskar. Approximation by superposition of sigmoidal and radial basis functions. *Adv. App. Math.*, 13:350–373, 1992.

[Mha96] H. N. Mhaskar. Neural networks for optimal approximation of smooth and analytic functions. *Neural Computation*, 8:164–177, 1996.

[MK94] M. Matsumoto and Y. Kurita. Twisted GFSR generators II. *ACM Trans. Modeling Comp. Syst.*, 4:254–266, 1994.

[MMR04] S. Mukhopadhyay, H. Mahmoodi, and K. Roy. Statistical design and optimization of SRAM cell for yield enhancement. In *Proc. IEEE/ACM Int. Conf. on CAD, 2004*.

[MMR05] H. Mahmoodi, S. Mukhopadhyay, and K. Roy. Estimation of delay variations due to random-dopant fluctuations in nanoscale CMOS circuits. *IEEE J. Solid-State Circuits*, 40(3):1787–1796, 2005.

[MP43] W. S. McCullough and W. Pitts. A logical calculus of the ideas immanent in nervous activity. *Null. Math. Biophys.*, 5:115–133, 1943.

[MTM97] C. Malthouse, A. C. Tamhane, and R. S. H. Mah. Nonlinear partial least squares. *Comp. Chem. Engg.*, 21(8):875–890, 1997.

[Nie78] H. Niederreiter. Quasi-Monte Carlo methods and pseudo-random numbers. *Bull. Amer. Math. Soc.*, 84(6):957–1041, 1978.

[Nie87] H. Niederreiter. Point sets and sequences with small discrepancy. *Monatsh. Math.*, 104(4):273–337, 1987.

[Nie88] H. Niederreiter. Low-discrepancy and low-dispersion sequences. *J. Number Theory*, 30:51–70, 1988.

[Nie92] H. Niederreiter. *Random Number Generation and Quasi-Monte Carlo Methods*. SIAM, Philadelphia, 1992.

[Nie98] H. Niederreiter. The algebraic geometric approach to low-discrepancy sequences. In H. Niederreiter, P. Hellekalek, G. Larcher, and P. Zinterhof, editors, *Monte Carlo and Quasi-Monte Carlo Methods 1996*, pages 139–160. Springer, New York, 1998.

[Nik50] S. M. Nikolskij. On the problem of approximation estimate by quadrature formulas. *Usp. Mat. Nauk*, 5:165–177, 1950 (in Russian).

[NT96a] S. Ninomiya and S. Tezuka. Toward real-time pricing of complex financial derivatives. *App. Math. Finance*, 3(1):1–20, 1996.

[NT96b] S. Ninomiya and S. Tezuka. Toward real-time pricing of complex financial derivatives. *App. Math. Finance*, 3(1):1–20, 1996.

[NW90] D. Nguyen and B. Widrow. Improving the learning speed of 2-layer neural networks by choosing initial values of the adaptive weights. In *Proc. Int. Joint Conf. Neural Networks, 1990.*

[NX96] H. Niederreiter and C. P. Xing. Low-discrepancy sequences and global function fields with many rational places. *Finite Fields Appl.*, 2:241–273, 1996.

[OE04] G. Ökten and W. Eastman. Randomized quasi-Monte Carlo methods in pricing securities. *J. Econ. Dyn. Control*, 28(12):2399–2426, 2004.

[Owe95] A. B. Owen. Randomly permuted (t, m, s)-nets and (t, s)-sequences. In H. Niederreiter and P. J.-S. Shiue, editors, *Monte Carlo and Quasi-Monte Carlo Methods in Scientific Computing*, pages 299–317. Springer, New York, 1995.

[Owe97a] A. B. Owen. Monte Carlo variance of scrambled net quadrature. *J. Numer. Anal.*, 34(5):1884–1910, 1997.

[Owe97b] A. B. Owen. Scrambled net variance for integrals of smooth functions. *Ann. Stats.*, 25(4):1541–1562, 1997.

[Owe98a] A. B. Owen. Latin supercube sampling for very high-dimensional simulations. *ACM Trans. Modeling Comp. Sim.*, 8(1):71–102, 1998.

[Owe98b] A. B. Owen. Scrambling Sobol' and Niederreiter–Xing points. *J. Complexity*, 14(4):466–489, 1998.

[Owe03a] A. B. Owen. Variance with alternative scramblings of digital nets. *ACM Trans. Modeling Comp. Sim.*, 13(4):363–378, 2003.

[Owe03b] A. B. Owen. The dimension distribution and quadrature test functions. *Stat. Sin.*, 13:1–17, 2003.

[PDW89] M. J. M. Pelgrom, A. C. J. Duinmaijer, and A. P. G. Welbers. Matching properties of MOS transistors. *IEEE J. Solid-State Circuits*, 24(5):1433–1440, 1989.

[Pet98] P. P. Petrushev. Approximation by ridge functions and neural networks. *SIAM J. Math. Anal.*, 30(1):155–189, 1998.

[PFTV92] W. H. Press, B. P. Flannery, A. A. Teukolsky, and W. T. Vetterling. *Numerical Recipes in C: The Art of Scientific Computing*, 2nd edition. Cambridge University Press, Cambridge, 1992.

[Pic75] J. Pickands III. Statistical inference using extreme order statistics. *Ann. Stats.*, 3(1):119–131, 1975.

[Pir02] G. Pirsic. A software implementation of Niederreiter–Xing sequences. In K.-T. Fang, F. J. Hickernell, and H. Niederreiter, editors, *Monte Carlo and Quasi-Monte Carlo Methods 2000*, pages 434–445. Springer, New York, 2002.

[Pra83] B. L. S. Prakasa Rao. *Nonparametric Functional Estimation.* Academic Press, New York, 1983.

References

[PT95]   S. Paskov and J. Traub. Faster valuation of financial derivatives. *J. Portfolio Management*, 22:113–120, 1995.

[PW72]   W. W. Peterson and E. J. Weldon, Jr. *Error-Correcting Codes*, 2nd edition. MIT Press, Cambridge, 1972.

[RB95]   J. Rifà and J. Borrell. A fast algorithm to compute irreducible and primitive polynomials in finite fields. *Theory Comput. Syst.*, 28(1):13–20, 1995.

[Ren03] M. Rencher. *What's Yield Got to Do with IC Design*. EETimes, Brussels, 2003.

[Res87] S. I. Resnick. *Extreme Values, Regular Variation and Point Processes*. Springer, New York, 1987.

[Rip96] B. Ripley. *Pattern Recognition and Neural Networks*. Cambridge University Press, Cambridge, 1996.

[Ros60] H. H. Rosenbrock. An automatic method for finding the greatest or least value of a function. *Computer J.*, 3:175–184, 1960.

[Rot80] K. F. Roth. On irregularities of distribution IV. *Acta Arith.*, 37:67–75, 1980.

[RSBS04] R. Rao, A. Srivastava, D. Blaauw, and D. Sylvester. Statistical analysis of subthreshold leakage current for VLSI circuits. *IEEE Trans. VLSI Syst.*, 12(2):131–139, 2004.

[RV98]   G. Reinsel and R. Velu. *Multivariate Reduced-Rank Regression, Theory and Applications*. Springer, Berlin, 1998.

[SC92]   X. Sun and E. W. Cheney. The fundamentality of sets of ridge functions. *Aequ. Math.*, 44:226–235, 1992.

[SK05]   I. M. Sobol' and S. S. Kucherenko. Global sensitivity indices for nonlinear mathematical models. Review. *Wilmott Magazine*, 2:2–7, 2005.

[SKC99] J. F. Swidzinski, M. Keramat, and K. Chang. A novel approach to efficient yield estimation for microwave integrated circuits. In *IEEE Midwest Symp. Circuit Syst., 1999*.

[Smi85] R. L. Smith. Maximum likelihood estimation in a class of non-regular cases. *Biometrika*, 72:67–92, 1985.

[Smi87] R. L. Smith. Estimating tails of probability distributions. *Ann. Stats.*, 15(3):1174–1207, 1987.

[Smo63] S. Smolyak. Quadrature and interpolation formulas for tensor products of certain classes of functions. *Dokl. Akad. Nauk SSSR*, 4:240–243, 1963.

[Sob67] I. M. Sobol'. The distribution of points in a cube and the approximate evaluation of integrals. *U.S.S.R. Comp. Math. and Math. Phys.*, 7(4):86–112, 1967 (English translation).

[Sob76] I. M. Sobol'. Uniformly distributed sequences with an additional uniform property. *U.S.S.R. Comp. Math. and Math. Phys.*, 16:1332–1337, 1976 (English translation).

[SP81]   K. Singhal and J. F. Pinel. Statistical design centering and tolerancing using parameter sampling. *IEEE Trans. Circuits Syst.*, 28(7):692–702, 1981.

[Spa95] J. Spanier. Quasi-Monte Carlo methods for particle transport problems. In H. Niederreiter and P. J.-S. Shiue, editors, *Monte Carlo and Quasi-Monte Carlo Methods in Scientific Computing*, pages 121–148. Springer, New York, 1995.

[SR07a] A. Singhee and R. A. Rutenbar. Beyond low-order statistical response surfaces: latent variable regression for efficient, highly nonlinear fitting. In *Proc. IEEE/ACM Design Autom. Conf., 2007*.

[SR07b] A. Singhee and R. A. Rutenbar. From finance to flip-flops: a study of fast quasi-Monte Carlo methods from computational finance applied to statistical circuit analysis. In *Proc. Int. Symp. Quality Electronic Design, 2007*.

[SR07c] A. Singhee and R. A. Rutenbar. Statistical Blockade: a novel method for very fast Monte Carlo simulation of rare circuit events, and its application. In *Proc. Design Autom. Test Europe, 2007*.

[Ste87] M. Stein. Large sample properties of simulations using Latin hypercube sampling. *Technometrics*, 29(2):143–151, 1987.

[Str71] A. H. Stroud. *Approximate Calculation of Multiple Integrals*. Prentice-Hall, Englewood Cliffs, 1971.

[SVK94] S. S. Sapatnekar, P. M. Vaidya, and S.-M. Kang. Convexity-based algorithms for design centering. *IEEE Trans. Computer-Aided Design*, 13(12):1536–1549, 1994.

[SWCR08] A. Singhee, J. Wang, B. H. Calhoun, and R. A. Rutenbar. Recursive Statistical Blockade: an enhanced technique for rare event simulation with application to SRAM circuit design. In *Proc. Int. Conf. VLSI Design, 2008*.

[Tez93] S. Tezuka. Polynomial arithmetic analogue of Halton sequences. *ACM Trans. Modeling Comp. Sim.*, 3(2):99–107, 1993.

[Tez95] S. Tezuka. *Uniform Random Numbers: Theory and Practice*. Kluwer Academic, New York, 1995.

[Tez05] S. Tezuka. On the necessity of low-effective dimension. *J. Complexity*, 21:710–721, 2005.

[Van35] J. G. Van der Corput. Verteilungsfunktionen. *Proc. Ned. Akad. v. Wet.*, 38:813–821, 1935 (in Dutch).

[VK61] B. A. Vostrecov and M. A. Kreines. Approximation of continuous functions by superpositions of plane waves. *Soviet Math. Dokl.*, 2:1326–1329, 1961.

[vM36] R. von Mises. La distribution de la plus grande de n valeurs. In *Selected Papers 2*, pages 271–294. American Mathematical Society, Providence, 1936.

[VRK$^+$04a] C. Visweswariah, K. Ravindran, K. Kalafala, S. G. Walker, and S. Narayan. First-order incremental block-based statistical timing analysis. In *Proc. IEEE/ACM Design Autom. Conf., June 2004*.

[War72] T. T. Warnock. Computational investigations of low discrepancy point sets. In S. K. Zaremba, editor, *Applications of Number Theory to Numerical Analysis*, pages 319–343. Academic Press, New York, 1972.

[WF03] X. Wang and K.-T. Fang. The effective dimension and quasi-Monte Carlo integration. *J. Complexity*, 19(2):101–124, 2003.

[WF05] I. H. Witten and E. Frank. *Data Mining: Practical Machine Learning Tools and Techniques*, 2nd edition. Morgan Kaufmann, San Francisco, 2005.

[Wo91] H. Woźniakowski. Average case complexity of multivariate integration. *Bull. Amer. Math. Soc.*, 24(1):185–194, 1991.

[WRWI84]	S. Wold, A. Ruhe, H. Wold, and W. J. Dunn, III. The collinearity problem in linear regression. The partial least squares (PLS) approach to generalized inverses. *J. Sci. Stat. Comput.*, 5(3):735–743, 1984.
[WS07]	X. Wang and I. H. Sloan. Low discrepancy sequences in high dimensions: how well are their projections distributed? *J. Comput. Appl. Math.*, 213(2):366–386, 2008.
[WSE01]	S. Wold, M. Sjöström, and L. Eriksson. PLS-regression: a basic tool of chemometrics. *Chemometr. Intell. Lab. Syst.*, 58:109–130, 2001.
[WSRC07]	J. Wang, A. Singhee, R. A. Rutenbar, and B. H. Calhoun. Modeling the minimum standby supply voltage of a full SRAM array. In *Proc. Europ. Solid State Cir. Conf., 2007.*
[XN95]	C. P. Xing and H. Niederreiter. A construction of low-discrepancy sequences using global function fields. *Acta Arith.*, 73:87–102, 1995.
[YKHT87]	T.-K. Yu, S. M. Kang, I. N. Hajj, and T. N. Trick. Statistical performance modeling and parametric yield estimation of MOS VLSI. *IEEE Trans. Computer-Aided Design*, 6(6):1013–1022, 1987.
[ZC06]	W. Zhao and Y. Cao. New generation of predictive technology model for sub-45 nm early design exploration. *IEEE Trans. Electron Devices*, 53(11):2816–2823, 2006.

Index

$C(X)$, 13
$L_p(X)$, 13
b-ary box, 73
p-norm, 13
(t, m, s)-net, 72, 73
(t, s)-sequence, 72, 74, 81
    digital, 79
    discrepancy, 75

A
acceptance region, 63
activation function, 11
ANOVA decomposition, 94, 175

B
balancing, 73, 92
bandgap voltage reference, 52, 114
Bayesian regularization, 37, 41
bias–variance tradeoff, 19
Black–Scholes model, 62
blockade filter, 143

C
causal dependency, 35
Central Limit Theorem, 129
characteristic function, 63, 68, 119
classification, 137
    linear, 137
classification threshold, 142
compact set, 13
conditional CDF, 127
conditionals, 156
confidence interval, 159
cross-validation, 37, 43
curse of dimensionality, 65

D
data retention voltage, 156
    distribution, 166
dense set, 13
digital method, 78
    Faure sequence, 80
    Niederreiter sequence, 80
    Niederreiter–Xing sequence, 81
    Sobol' sequence, 80
digital net, 79
digital sequence, 78
digital (t, s)-sequence, 79
direction number, 83, 86
discrepancy, 68, 69
    Faure sequence, 75
    L_2 star discrepancy, 70
    random sequence, 70
    Sobol' sequence, 75
    star discrepancy, 68, 69
    (t, s)-sequence, 75
disjoint tail regions, 156, 157
dropout voltage
    bandgap voltage reference, 53
Dutch dikes, 125

E
effective dimension, 95, 97, 100, 101
    superposition, 95
    truncation, 95
expectation, 22
extreme value theory, 125, 128
extremely rare events, 159

F
Faure sequence, 75
    digital method, 80
    discrepancy, 75

A. Singhee, R.A. Rutenbar, *Novel Algorithms for Fast Statistical Analysis of Scaled Circuits*, Lecture Notes in Electrical Engineering 46,
© Springer Science + Business Media B.V. 2009

Fisher–Tippett, 128
Fréchet, 128

G
Gauss–Newton method, 40
generalization, 21, 36
generalized extreme value, 129
generalized Pareto distribution, 131
generator matrix, 79
global sensitivity, 34
global sensitivity index, 95
gradient, 39
Gray code, 88
Gumbel, 128

H
Halton sequence, 77
Hardy and Krause, variation, 71
Hessian, 39, 42
high replication circuit, 123, 173
homogeneous polynomial, 15
hyperbolic tangent, 28

I
input-referred correlation, 35, 50
integration error, 65
    estimate, 103
        quasi-Monte Carlo, 104
integration lattice, 72
IRC, see input-referred correlation

J
Jacobian, 40

K
kernel trick, 9
Koksma–Hlawka, 69, 96
Kronecker product, 5

L
latent variable, 8, 19, 29
latent variable regression, 8
Latin hypercube sampling, 88
    construction, 89
        scrambled (t, m, s)-net, 91
        Sobol' sequence, comparison with, 98, 111
    variance, 90, 98, 110
Latin supercube sampling, 121
LDS, see low-discrepancy sequence
Levenberg–Marquardt, 37, 38, 40, 42
LHS, see Latin hypercube sampling
likelihood, 134
linear model, 4
linear projection, 29
Lipschitz condition, 105

log-likelihood function, 134
logistic function, 28
low-discrepancy sequence, 71, 72
low-rank approximation, 6

M
master–slave flip-flop, 45, 114, 153
maximum domain of attraction, 129, 130
    tail regularity, 131
maximum likelihood estimation, 134
    variance, 135
MDA, see maximum domain of attraction
mean excess function, 161
measure, probability, 22
mixture importance sampling, 124
moment matching, 135
Monte Carlo, 66
    convergence, 66, 69, 119
        Bakholov, 66
    variance, 67, 88

N
neural network, 11
Newton's method, 39
Niederreiter sequence
    digital method, 80
Niederreiter–Xing sequence
    digital method, 81

O
option, 61
    Asian option, 61
    strike price, 61
overfitting, 20, 33

P
peaks over threshold, 128
perceptron, 11
PPR, see projection pursuit
primitive polynomial, 83, 86
probability-weighted distribution
    variance, 136
probability-weighted moments, 135
PROBE, 6
projection matrix, 8
projection pursuit, 10, 12, 18
    convergence, 21
    Hall, 27
    Huber, 24, 26
    Jones, 26
projection vector, 8, 19
projection weight, 8, 34

Q
quadratic model, 5
quadrature, 65

Index

quasi-Monte Carlo, 72
    circuits, 101
    convergence, 119
    patterns, 92
    skip initial points, 103

R

radical inverse function, 77
random dopant fluctuation, 45
rank, 37
rare events, 127
reduced rank regression, 8
regular variation of function, 132
regularization, 41
relative global sensitivity, 34
residue, 18, 22
response surface model, 4
ridge function, 10
    degree of approximation, 16
        Maiorov, 17
    density, 14
        Sun and Cheney, 15
        Vostrecov and Kreines, 15
    Fourier series, 12
roughness penalty, 41

S

sample maximum, 128
    limiting distribution, 128
sample mean excess plot, 161
scrambled sequence, 105
    linear matrix scrambling, 107
    Owen's method, 106
    Sobol', 108
    variance, 105
scrambling, 90
separating hyperplane, 139
    optimal, 140
sigmoid, 28
    derivative, 29
SiLVR, 27, 29
    algorithm, 31
    comparison with PROBE, 55
    complexity, 31
    convergence, 31
        Barron, 32

    Chui and Li, 32
    Cybenko, 32
    objective, 30
    overfitting, 33
slowly varying function, 132
smooth, 18
smoothing, 122
Sobol' sequence, 75, 82
    construction, 82
    digital method, 80
    discrepancy, 75
    Latin hypercube sampling, comparison with, 98, 111
    properties A and A', 87
    scrambling, 108
Spearman's rank correlation, 37, 102, 115, 151
SRAM, 114, 123, 147, 149
statistical blockade, 125, 143, 144
    comparison, 148, 152, 155, 168
    recursive formulation, 163–165
    variance, 160
steepest descent, 39
Stone–Weierstrass theorem, 14
support points, 141
support vector, 141
support vector machine, 138

T

tail, 127
    fitting, 133
    heavy, 126, 153
    limiting distribution, 130
tail threshold, 127
two-stage opamp, 47

V

Van der Corput sequence, 76
variable-dimension mapping, 101
variance reduction, 90

W

Weibull, 128
Weierstrass theorem, 13
Wiener process, 62

Y

yield, circuit, 64